The World Book of Space Exploration

WONDERS

OF THE

UNIVERSE

The World Book of Space Exploration

WONDERS OF THE

UNIVERSE

World Book, Inc.
a Scott Fetzer company
Chicago

In this volume, the
sizes of and distances
between the planets,
sun, and moon are
not always drawn to
scale. This is because of
the large sizes and
tremendous distances
involved.

Printed in the United States of America.
ISBN 0-7166-3229-2
Library of Congress Catalog
Card No. 90-70865

c/ij

Staff

President
Peter Mollman

Publisher
William H. Nault

Editorial
Editor in chief
Robert O. Zeleny

Executive editor
Dominic J. Miccolis

Associate editor
Maureen Mostyn Liebenson

Senior editor
Karen Zack Ingebretsen

Contributing editors
Waldemar Bojczuk
Kathy Klein

Permissions editor
Janet T. Peterson

Art
Art director
Roberta Dimmer

Assistant art director
Joe Gound

Photography director
John S. Marshall

Book design
Chestnut House

Product production
Director, manufacturing
Henry Koval

Manager, manufacturing
Sandra Van den Broucke

Assistant manager, manufacturing
Eva Bostedor

Director, pre-press services
Jerry Stack

Product managers
Randi Park
Joann Seastrom

Proofreaders
Anne Dillon
Marguerite Hoye
Daniel J. Marotta

Contents

THE SKY ABOVE, THE EARTH BELOW

This Mesopotamian boundary stone of 1100 B.C. shows the moon at the top between its "children"—Venus (left) and the sun (right).

Lightning splits the night sky over Kitt Peak National Observatory in Arizona.

Since the dawn of the human race, the night sky has been the object of much study and speculation. Persons from all walks of life have tried to explain the wonders that appear night after night in the sky. This section reviews some of the major theories that have attempted to explain our place in the universe. It also introduces some of the important stargazers who made major discoveries about the sky above and the earth below.

A portion of the Bayeux Tapestry, created in France in the Middle Ages, shows King Harold of England being told of a bright comet in 1066.

THE NIGHT SKY

Go out on a clear, moonless night into the country, far from the air pollution and lights of any city or town, lie down on the ground, and look up. The sight is overwhelming. Stars by the hundreds fill the heavens. Some brilliant and some faint, they dot the black sky from one horizon to the other.

Most of us don't see the night sky in its true splendor, because we live in places where artificial lights illuminate the sky and spoil the view. But from a truly dark, rural location, our view of the heavens is essentially the same that every age of the human race has beheld. Stone Age men and women looked up at this sight from their campfires. Shepherds in ancient Mesopotamia and Egypt marveled at the same celestial view we see today.

The ancient monument of Stonehenge in England was built sometime between 2800 and 2000 B.C. It may have served as an astronomical calendar to predict the seasons of the year and even eclipses of the sun and moon.

Today, we know this impression is completely wrong. Earth is round, not flat. It is a tiny speck adrift in a universe so huge that no one can imagine its size. The stars are huge balls of glowing gas—trillions and trillions of them, unbelievably far away. Many of them probably have planets of their own. Space is full of other wonders so strange, so bizarre, so beautiful and unimaginable that no one could possibly have dreamed them up before they were discovered.

But what we know about the universe has taken ages of slow detective work. There were many false starts, wrong turns— and a few cowardly retreats.

ANCIENT IDEAS

The earliest peoples thought, naturally enough, that the universe was only what they saw: a sky dome over a flat Earth. Beyond that, they simply made up stories. The ancient Babylonians believed the world floated on an immense ocean, while the gods lived on top of the sky dome. The sun came through a door in the sky every morning and left through another door in the evening.

What are the stars? How big is the universe? Where do we fit in? People have been asking these questions since earliest times, but the answers have been slow in coming.

At first, people could judge only from what they saw: a great black dome of sky arching above a flat Earth, like a bowl turned upside down on a tabletop, with the people inside the bowl. This is still the way the universe looks: simple, small, self-contained.

The Egyptians saw the universe as a room, with the sky as the ceiling and Earth as the floor. Egypt, of course, was in the center. The stars hung from the sky-ceiling like lamps, and the sun god Ra sailed around the sky once a day in a boat.

As the ancient Hindus imagined it, the flat Earth rested on the backs of four elephants, who were standing on the back of a giant turtle, which swam in a river of milk.

Such ideas must have seemed perfectly logical to the people who believed them. They would have scoffed at the notion that Earth is round and moves through space. But there was one big difference between their ideas and those that would follow. Their myths were not based on any evidence. They were simply made up. The idea of testing such ideas to see whether or not they are true never seems to have occurred to the people of ancient Mesopotamia and Egypt.

The ancient astronomer-priests did know much practical astronomy. They could forecast eclipses and tell just what time

Many ancient cultures worshiped the sun. Here, the Egyptian pharaoh Akhenaton, who ruled Egypt from about 1367 to 1350 B.C., is shown worshiping the sun god, Aton.

12

of year certain stars would appear at dawn. But they did not understand the reasons for the things they observed in the sky. They thought of the celestial bodies as gods, or as whimsical, chaotic forces that no one could understand. So there seemed to be no point in trying to find out why these bodies behaved as they did.

GREECE AND THE BIRTH OF SCIENCE

About 2,500 years ago, in the land of Greece, a new way of thinking began to take root. In the broadest terms, it was the idea that the world is knowable—that the universe acts in predictable ways according to natural laws.

The Greek philosophers developed and refined geometry and much of mathematics, and they were amazed to discover the hidden laws that numbers obey. Pythagoras believed that the secret laws of numbers were so holy and perfect they should be kept hidden from ordinary people. One of Pythagoras' beliefs was that Earth was round. Other philosophers, studying the motions of the heavenly bodies, proposed various ideas for why the sun, moon, stars, and planets move as they do in the sky. They applied their mathematics to the problem, and some came to believe that the celestial bodies are carried around the sky on invisible, hollow spheres, with Earth at the center.

This might not seem like much improvement on sun boats, elephants, and turtles. But there was a big difference. The Greeks' universe was not run by the whims of irrational gods; it was orderly like a piece of machinery. Its motions could be studied and predicted.

The first person to accurately measure the size of Earth was a scholar named Eratosthenes who lived in the Egyptian city of Alexandria, then part of the Greek Empire. He knew that in the city of Syene, farther south, the sun stood straight overhead at noon on the day of the summer solstice, so that pillars cast no shadows and sunlight shone to the bottoms of wells. But on the same day in Alexandria, the sun was not quite overhead, and a vertical post cast a shadow. Measurement showed that here, the sunlight fell at an angle of about 7° from the vertical—or about one-fiftieth of a circle. Believing in the round-earth theory proposed by Pythagoras, Eratosthenes set out to determine the size of Earth.

He hired a man to pace out the distance between the two cities. It came to about 500

Claudius Ptolemaeus, better known as Ptolemy, was one of the greatest astronomers and geographers of ancient times. He made his astronomical observations at Alexandria, Egypt, about A.D. 150.

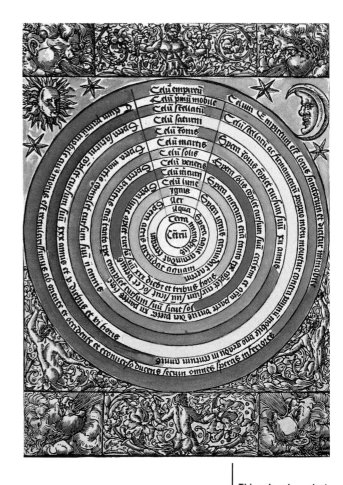

This colored woodcut from the 1500's shows the Ptolemaic universe, with Earth in the center, and the sun, moon, and other planets revolving around it.

miles (800 kilometers). Since he knew this distance had to be one-fiftieth of Earth's circumference, then Earth must be about 25,000 miles (40,000 kilometers) around.

He was right. Simply by measuring the angle formed by shadows, he had calculated the size of Earth.

For anyone interested in the mysteries of nature, this was an exciting time to be alive. Other philosophers realized Earth is not motionless but rotates, or turns on its axis, once a day; this is why the sun, moon, and stars appear to rise and set. When we see the sun go down in the west, we are actually seeing the western horizon rise. The rising and setting is not due to any motion of the sun.

Aristarchus, another Greek, went further. Knowing the size and shape of Earth, he managed to estimate the distance to the moon. By studying the phases of the moon, he deduced that the sun is much farther away, which

meant it had to be bigger than Earth. He then proposed that the planets—"wandering stars" seen among the constellations—circle the sun. He declared that Earth itself is a planet and does the same.

He got it entirely right. Human thought had gone from a flat Earth and dome sky to the correct picture of our spinning planet flying through space—and all about 1,900 years before the invention of the telescope.

Most other thinkers of the time rejected Aristarchus' idea as too strange to be true. But the

14

Nicolaus Copernicus (1473–1543) was a Polish astronomer who developed the theory that Earth is a moving planet and not the center of the solar system. He is considered the founder of present-day astronomy.

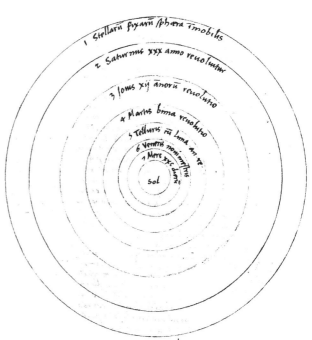

This illustration from Copernicus' Concerning the Revolutions of the Celestial Spheres (1543) shows Earth and the other planets revolving around the sun. This sun-centered theory revolutionized astronomy.

evidence was there. In time all would probably have come to believe it. Many other discoveries were being made in other areas of science—geography, physics, medicine, mathematics. There is no telling where it all might have led. But ancient Greek science had a fatal flaw. For the most part, the philosophers did not believe in getting involved with the practical world. They believed science and mathematics were too pure and holy to be taught to common people. But the people who built things— who did the work and could have put new discoveries to use—*were* the common people. So, much of Greek science was stillborn. It died out in the wars, tyranny, and chaos that finally destroyed classical civilization. In the centuries that followed, the discoveries of the Greeks were all but forgotten.

One idea of the Greeks, though, did arouse interest during the Middle Ages. Plato and Aristotle taught that all was divided into two parts: the pure, perfect, and eternal celestial regions above, and the dirty, flawed, changing Earth below. This division of sky from Earth seemed to match Christian teachings about divisions between perfect heaven versus fallen humanity, pure soul versus impure body. So the ideas of Aristotle were widely discussed and were accepted by many scholars. But church leaders put an end to inquiry into what astronomical bodies really are. The flat Earth returned, and for a long time it was even forbidden to teach that Earth is round.

THE COPERNICAN REVOLUTION

Astronomy did not stop during this time, however. There was much interest in predicting where the planets would be among the constellations on any given day. The best system for doing this was a mathematical description of the paths along which the sun, the moon, and each of the planets were supposed to orbit the Earth, developed around A.D. 150 by an Egyptian astronomer named Ptolemy.

When some people in the late Middle Ages tried to picture the

15

solar system as Ptolemy described it, they imagined that an arrangement of invisible crystal spheres carried the celestial bodies through the sky. They believed that the moon, for example, was attached to a small, turning sphere, which in turn rode upon a large, rotating sphere that encircled the Earth. The sun and each of the planets also traveled through the heavens on its own set of rotating crystal spheres. Beyond the spheres of the planets was one that carried the stars. Earth was just off the center of the series of spheres that surrounded it. This system was complicated and awkward, but it worked fairly well as a mathematical tool for solving the motions of the planets.

However, during the mid- to late Middle Ages, some men who carefully observed the planets found that the Ptolemaic system did not work perfectly. And some people had a hard time believing the sky was really jammed full of invisible machinery. Eventually, a few even began to suggest that a way should be found to describe the motions of the planets more ac-

curately. One such person was Nicolaus Copernicus, who lived during the early 1500's. Copernicus was a canon of the Roman Catholic Church in what is now Poland. He spent much of his life trying to find a more philosophically and aesthetically satisfactory way than Ptolemy's theory to account for the motions of the planets. He succeeded—by assuming that the planets, including Earth, travel around the sun. The great theory of Aristarchus had been rediscovered.

Copernicus was a timid, reclusive man. He did not publish his theory until shortly before his death. Although Copernicus' theory received little attention at first, it eventually raised stormy debates. This was because many traditional academics and clergy believed deeply in a motionless Earth at the center of the universe, and in Aristotle's perfect, invisible spheres filling the skies with perfect circular motion.

A lively participant in those debates was Galileo Galilei, an outspoken and popular teacher of mathematics in Italy. In 1609, Galileo first looked at the sky with the newly invented telescope. With it, he saw that the moon has mountains and plains like those on Earth—though followers of Aristotle taught that the moon was a smooth, perfect sphere. He discovered four tiny moons orbiting the planet Jupiter—though traditionalists who believed in perfection in the heavens denied that such things could exist. Some even refused to look. He discovered that the planet Venus shows phases like those of the moon, which he saw as evidence that Venus circles the sun.

For such reasons, Galileo became a champion of the Copernican theory. He was a sharp-tongued debater and made many enemies. Some of them went to church authorities and urged

Galileo Galilei (1564–1642) has been called by some historians the founder of modern experimental science. Among other innovations, he made the first practical use of the telescope to discover many new facts about the moon and planets.

Johannes Kepler (1571–1630), a German astronomer and mathematician, discovered three important laws of planetary motion, including the fact that every planet moves around the sun in an ellipse, not in a circle.

Two of Galileo's telescopes are on display in the Museum of the History of Science in Florence, Italy.

that Galileo be silenced, on the grounds that a moving Earth contradicts the Bible. It was a deadly accusation to make. A few years before, the Catholic Inquisition had sentenced Giordano Bruno, a Dominican priest and freethinker, to be burned alive at the stake, partly for teaching the Copernican theory.

The officials of the Inquisition summoned Galileo. They ordered him to abandon the "absurd and heretical" notion that the sun is stationary and Earth moves around it. Galileo agreed not to teach the Copernican theory as being true. A few years later, however, he published a book that clearly defended it. He was summoned back to the Inquisition and forced, under threat of torture, to deny formally and publicly that Earth goes around the sun. For the rest of his life, he was kept under house arrest, constantly guarded and watched.

But the Copernican theory could not be suppressed. It explained the behavior of the planets too well. And it did away with some of the imaginary celestial clockwork of Ptolemy.

Tycho Brahe, a Danish nobleman, spent many years observing the positions of planets and measuring planetary positions in relation to Earth and the sun. These measurements were far more accurate than any that had been recorded before. Using Brahe's data, Johannes Kepler, a teacher in Germany, greatly improved the Copernican system. After years of wrestling with the mathematics, he figured out that Earth and the planets move around the sun not in circles, but in ellipses, or oval-shaped paths. Kepler also discovered two mathematical principles that describe the speed of the planets in their orbits. With Kepler's three laws of planetary motion, the shape and speed of the planetary orbits were at last understood.

In the process, our true place in the solar system had been revealed. We are not at the center; rather, we are on a relatively small planet that revolves around the sun.

EARTH AND SKY REUNITED

The story is told that Isaac Newton, an English scientist, was sitting under an apple tree. An apple fell to the ground, and he suddenly realized that the force that pulls the apple to Earth also holds the moon in or-

Danish astronomer Tycho Brahe (1546–1601) developed a systematic approach for observing the planets and stars. He is pictured here in an observatory he built on the Danish island of Ven.

18

This large telescope was constructed in the seventeenth century by the Polish astronomer and instrument maker Johannes Hevelius. He built the world's leading observatory in Gdańsk, Poland. It was destroyed by fire in 1679.

bit. But, he wondered, what laws govern the effect of gravity? How does gravity work?

Newton searched for an explanation for years, and his solution was his great law of gravitation, a very simple but far-reaching notion: every object pulls every other object toward it by a gravitational force. The more massive an object, the stronger its gravity; the farther away it is, the weaker is gravity's effect. Earth is by far the most massive object near us, so it exerts the strongest force on us.

The moon is indeed subject to Earth's gravity, but instead of falling down, Newton showed, it must circle Earth in an orbit. It is, in effect, falling around Earth.

To picture how an orbit works, imagine yourself holding one end of a string, which has a ball tied to the opposite end. Raise your arm sideways and twirl your hand in a circular motion so that the ball "orbits" around your hand. You now have the "moon" orbiting around the "earth." Just as the string keeps the ball orbiting around your hand, so does gravity keep the moon orbiting around

Sir Isaac Newton (1642–1727), an English scientist, astronomer, and mathematician, showed through his theory of gravitation how the universe is held together.

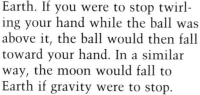

This huge reflecting telescope, nicknamed the "Leviathan," was built in 1845 by Lord Rosse, an Irish astronomer. It had a 72-inch (180-centimeter) mirror made of metal.

Earth. If you were to stop twirling your hand while the ball was above it, the ball would then fall toward your hand. In a similar way, the moon would fall to Earth if gravity were to stop.

When Newton applied his mathematical formula of gravity to the planets, he found that it almost exactly accounted for the planets' motions and the elliptical shapes of their orbits. All the workings of the solar system were explained by the same, simple law that governs a falling stone here on Earth.

Thus, Newton not only laid the foundations for much of modern physics, he overturned the doctrine of Aristotle that celestial bodies are some different, more ideal order of being than things on Earth. The objects in the sky have the same physical nature and obey the same physical law as things on the ground. Or, to look at it another way, we ourselves are made of celestial material. Our ordinary, everyday world is part of the cosmos. The celestial regions were once again fully open to inquiry and discovery.

THE GROWTH OF MODERN ASTRONOMY

By the time Newton died in 1727, astronomy and many other sciences were making great advances in Europe. Each new discovery became the foundation from which others could be made. Newton himself said that if he saw farther than anyone had before, it was because he stood on the shoulders of giants. His own work became the base for other astronomers to stand on and see farther still.

In all the time the planets had been studied, the stars were little more than background scenery. They were clearly farther away than the planets, but not much else could be told about them. Once Earth was known to go around the sun, however, a method for finding their distance became available.

The motion of Earth from one side of its orbit to the other should make nearby stars seem to move back and forth with respect to farther ones. You can

Bigger and better telescopes are constantly being pointed toward the sky. This large optical telescope is located at Kitt Peak National Observatory in Arizona.

THE REALM OF THE SUN

The sun is the center of our part of the universe. It is the main source of energy and the sustainer of life. Traveling around it are many interesting and unusual objects, including planets, asteroids, and comets. These objects each have their own unique features—from a volcano three times the height of Mount Everest, to a canyon that would stretch the length of the United States, to an entire world made of ice and smooth as a billiard ball.

The sun, planets, moons, and other objects that we call the solar system are the subject of this section. They are truly a realm of the fantastic.

On the right is a photograph of Earth as seen from space. On the far right is the crater Yuty on Mars, as photographed by the Viking 1 space probe.

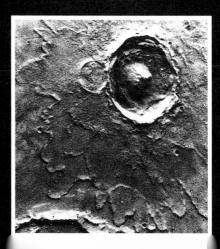

This picture of the solar corona was photographed in Bolivia during an eclipse of the sun on November 12, 1966.

In this color-coded image of Comet West, each color corresponds to a different level of light intensity.

The Voyager 2 space probe took this photograph of Saturn and its rings.

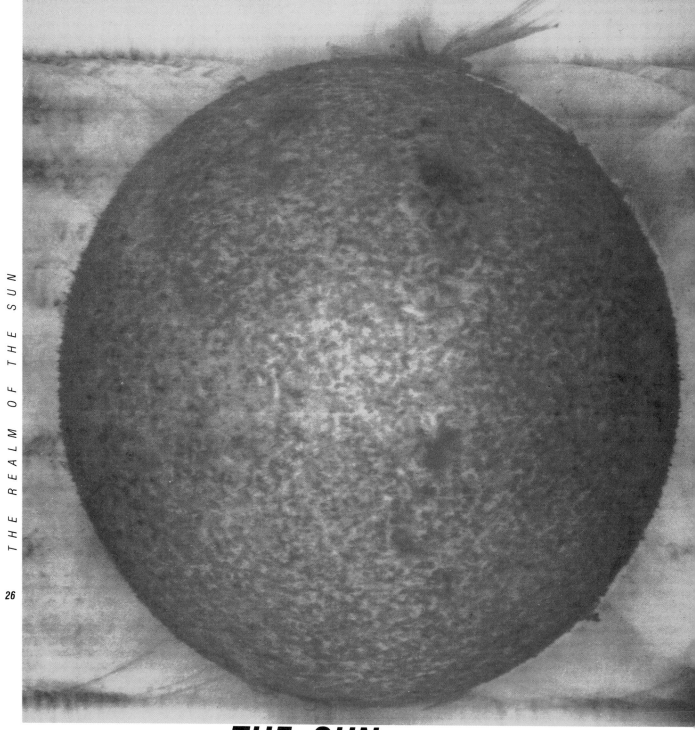

THE SUN

We live our lives under the influence of a glowing ball of gas, whose energy lights our days, gives us life-sustaining warmth, and nourishes all living things. The star we call the sun is an average star as stars go, just approaching middle age in its 10-billion-year life span. But to us, it is the most important star in the universe. It is also the nearest, and for that reason, the one that we know the most about.

From Earth, the sun is a brilliant disk, too bright to look at. Yet it is some 93 million miles (158 million kilometers) away, so far that its light takes more than 8 minutes and 20 seconds to reach us.

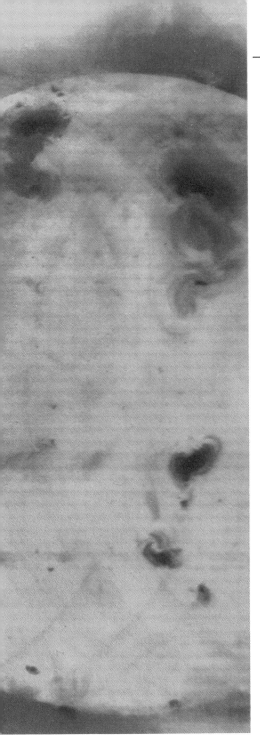

From here the sun looks small, but it is actually about 865,000 miles (1,384,000 kilometers) across. Like most stars, it is made mostly of hydrogen (about 75 percent of the sun's mass), with a smaller amount of helium (about 24 percent), and much smaller quantities of oxygen, carbon, nitrogen, and other elements.

THE SUN'S NUCLEAR FIRES

The sun produces an incredible amount of energy. About 30 minutes worth of the radiation Earth receives from the sun is enough to equal all the energy produced by every city and town on Earth in a year; and yet we receive only one two-billionths of the sun's total energy output. What could produce such enormous quantities of energy? The answer is nuclear fusion, the same process responsible for the awesome destructive power of a hydrogen bomb.

The sun's energy is unleashed deep within its interior, in its core. There, hydrogen atoms collide and fuse together to form helium atoms, and this fusion process produces the sun's energy. Because the core is hidden beneath other layers of gas thousands of miles thick, astronomers cannot directly observe it. But with the aid of sophisticated instruments and computers, they can calculate what it must be like. It is certainly extremely hot—probably 27,000,000° F. (15,000,000° C). The gas in the core is so highly compressed by the weight of the overlying layers that it is estimated to be 15 times more dense than lead—hardly like a gas at all.

If it were not so hot, the sun would surely collapse under its own gravity. But the extreme heat within the sun creates very high pressures that push outward, counteracting gravity.

The intense energy produced by nuclear fusion in the sun's core travels outward into the surrounding gas. Through most of the sun, energy travels by radiation—the same process that makes your face feel warm when you look through the window of a hot oven.

But in the upper layers, the energy travels by convection, the same process by which hot water in a tub cools off. In the sun's convection zone, gas heats up until it expands to form bubbles called granules. Because these are less dense—and therefore lighter—than their surroundings, the granules rise toward the sun's surface layers until they reach a layer called the photosphere, where they radiate energy, cool off, and sink back into deeper layers again.

The energy carried up by the granules is released in the photosphere in the form of sunlight. Although it contains all the colors of the rainbow, sunlight is predominantly yellow. Scientists know that the color of a star depends on its temperature. The photosphere's yellow color corresponds to a temperature of about 10,000° F. (5,500° C).

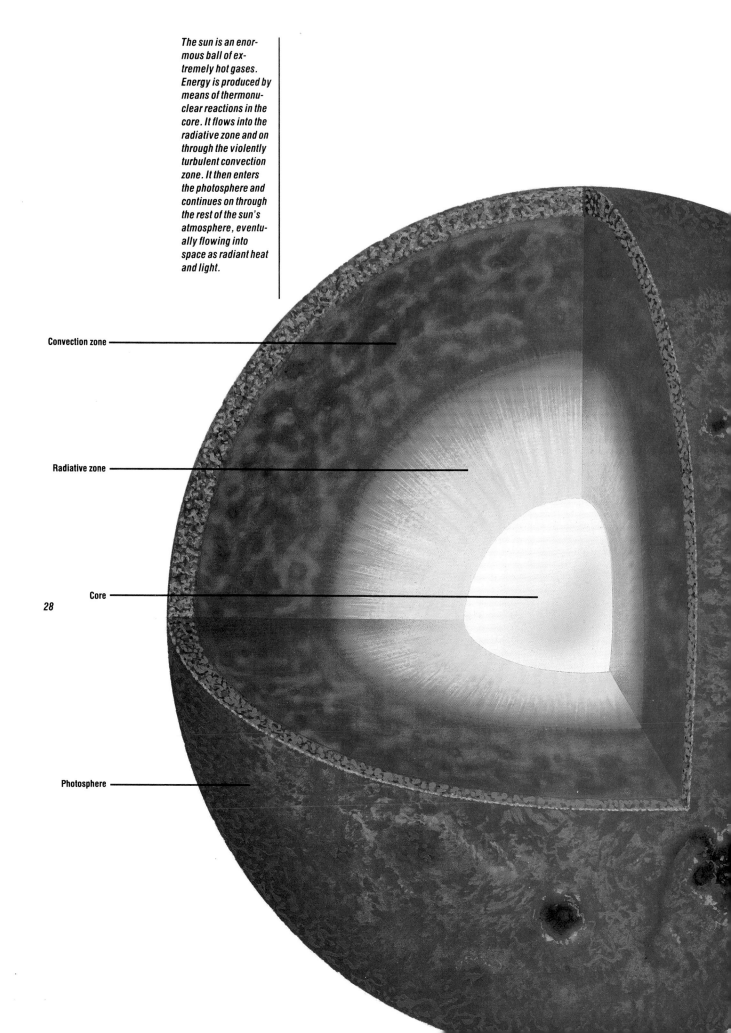

The sun is an enormous ball of extremely hot gases. Energy is produced by means of thermonuclear reactions in the core. It flows into the radiative zone and on through the violently turbulent convection zone. It then enters the photosphere and continues on through the rest of the sun's atmosphere, eventually flowing into space as radiant heat and light.

Convection zone

Radiative zone

Core

Photosphere

28

THE SUN'S SURFACE

The photosphere is the visible surface layer of the sun. It is not, of course, a solid surface. It is a thin layer, only about 340 miles (544 kilometers) thick. On astronomers' photographs, the photosphere has a coarse, grainy appearance resembling boiling oatmeal. And in fact, the grains are really the tops of the granules rising from the interior of the sun. Astronomers call this texture granulation.

Above the photosphere is a sparse, transparent layer called the chromosphere. It can be seen by the eye only during a total eclipse of the sun. Then it forms a thin, flame-red ring around the black shadow of the moon. The chromosphere is much hotter than the photosphere below, reaching temperatures of 50,000° F. (27,800° C) or more in its tenuous upper reaches.

Beyond the chromosphere is the corona, where temperatures reach some 4,000,000° F. (2,200,000° C). This jump in temperature between the chromosphere and corona occurs in a transition region only a few tens of miles (kilometers) across.

The corona is the outermost layer of the sun's atmosphere, stretching millions of miles (kilometers) into space. During a total solar eclipse, the corona may look like the petals of a pale white flower surrounding the blackened sun and the fiery red chromosphere.

SUNSPOTS AND THE SOLAR CYCLE

The photosphere is often dotted with markings called sunspots. Sunspots look dark compared to the bright photosphere because they are some 4,000° F. (2,200° C) cooler, yet they are still as hot as an acetylene torch. A typical sunspot might be 6,000 miles (10,000 kilometers) across—almost large enough to swallow Earth. But sunspots five times that size have also been recorded.

Sunspots show that the sun, like Earth, rotates. Photographs taken hours or days apart show sunspots moving across the solar disk. It takes about 25 earth-days for the sun's equator to make one full rotation, but the sun's poles are slower: they take 35 or 36 earth-days to go around. In this way, gaseous bodies like the sun are different from solid bodies, such as Earth, all parts of which rotate at the same rate at all latitudes.

The number of sunspots visible on the sun varies. When many sunspots are present, solar activity is on the increase. When there are few sunspots, the sun is relatively quiet. Astronomers have found that every 11 years the number of sunspots reaches a maximum. These sunspot maxima are the times when the sun is at its peak of activity. Roughly five and one-half years after each sunspot maximum, there is a sunspot minimum, and the sun is at its least active. Then, five and one-half years after that, another maximum occurs. The whole 11-year progression is called the sunspot cycle. 29

Some scientists have found evidence that climate on Earth may be influenced by the solar cycle. During most of the 1600's for example, there were far fewer sunspots on the sun than usual. Some astronomers believe this may have been the reason for the unusually cold weather that persisted during those years.

Further studies may reveal whether the Earth's weather patterns are affected by the solar events of an individual sunspot cycle or by solar variations occurring over a much longer period of time. In 1980, the United

The McMath Solar Telescope at Kitt Peak National Observatory in Arizona, the world's largest solar telescope, is operated by the National Solar Observatory.

States launched a spacecraft called *Solar Maximum*, or *Solar Max*, at the peak of the then-current sunspot cycle. The mission showed that reductions in the amount of solar energy that reaches the Earth's atmosphere correspond to the area of the sun covered by sunspots. The satellite continued to study the sun until 1984 when it malfunctioned. The *Challenger* space shuttle rescue mission successfully repaired *Solar Max*. However, the satellite re-entered the Earth's atmosphere on December 3, 1989, and burned up over the Indian Ocean.

What keeps sunspots cool? Scientists don't know for sure. But one possible answer is that convection of heat from the interior does not occur within sunspots. Instead of bubbling up and turning over, gas within a sunspot is trapped by the sun's strong magnetic field. Solar gas is so hot that the gas molecules bang into each other with enough force to strip away the outer electrons surrounding individual atoms. Such ionized gas cannot move wherever it wants, but rather only along magnetic field lines.

SOLAR PROMINENCES

Observations of the edge of the sun, using special filters, show flamelike clouds of glowing gas extending up into the sun's co-rona. These clouds, which are called prominences, can linger for periods from a few hours to days. Though prominences are most easily observed at the edge of the sun, they can occur anywhere on the surface. Sometimes a prominence will erupt away from the sun in a spectacular blast.

OUTBURSTS FROM THE SUN

The sun's steady shining is accompanied by other violent bursts of activity. On occasion, a brilliant outburst called a flare erupts in the chromosphere. A major flare can explode with the power of a 2 billion megaton bomb. When radiation from the flare reaches Earth eight minutes later, it can play havoc with radio transmissions. Protons and

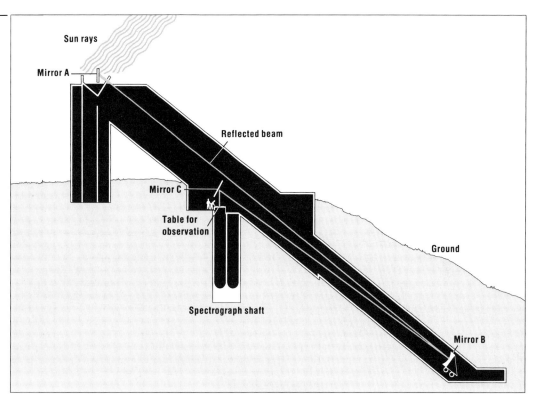

This cutaway view of the McMath Solar Telescope on Kitt Peak shows how it works. Sunlight striking Mirror A is reflected down the diagonal shaft. Mirror B, an image-forming mirror, sends the light back up the shaft. Mirror C finally reflects the image onto a table for observation.

Sun rays

Mirror A

Reflected beam

Mirror C

Table for observation

Ground

Spectrograph shaft

Mirror B

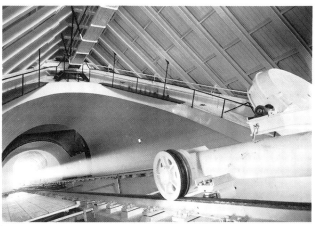

This is a view of the light path within the inclined tunnel of the McMath Solar Telescope.

electrons spewed out by the flare reach Earth in about 36 hours and may enter our upper atmosphere, where they create an aurora. For astronauts in space, the radiation from a large flare is a great hazard.

Flares are most common during times of sunspot maximum. Scientists are still trying to find out why flares occur and how their enormous energy is released. Current theories suggest that flares make use of energy stored in the sun's magnetic field.

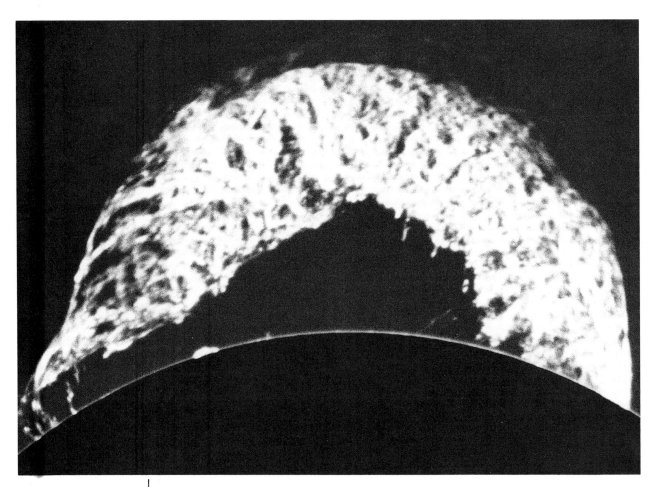

Shown in its early stages, the largest solar prominence ever observed became almost as large as the sun itself within one hour. After a few hours, it had disappeared completely.

The aurora borealis, or northern lights, is a natural display of light, caused by the interaction of the solar wind with Earth's upper atmosphere.

The sun's corona evaporates away in a stream of charged particles called the solar wind. The solar wind pervades the solar system to its very edge, where it merges with winds from other stars.

THE DEATH OF THE SUN

Like all stars, the sun will not last forever. The sun is burning its hydrogen at a rate of 11 billion pounds (5 billion kilograms) per second. But even at that rate, it has several billion years left before nearly all of the hydrogen in its core is converted into helium. Astronomers believe that as it uses up hydrogen, the sun will grow in size and brightness. Five and one-half billion years from now, the sun may be 1.4 times its present size and about twice its present brightness. All the hydrogen in the core will be gone; only the helium "ashes" of its nuclear fires will remain.

At this point, not enough energy will be produced in the sun's core to counteract the star's own gravity. As the fires in the core die out, the core will begin to contract. But the collapse will generate enough heat to burn the thin shell of leftover hydrogen around the core. This new fire will cause the star's surface layers to expand. Astronomers calculate that after an additional 1.5 billion years, the sun will be 3.3 times larger than it is today, and 250 million years after that, it will swell to 100 times its present size—all the way out to the innermost planet, Mercury. It could be 500 times brighter than it is today, what astronomers call a red giant. The intense heat from the enormous, deep-red sun will turn Earth's surface into a sea of molten rock.

After a brief but fiery career as a red giant (lasting perhaps 250 million years), the sun will contract again as the very last of its hydrogen is converted to helium. Temperatures in the core will rise as the sun falls in on itself, and when the core heats to 180,000,000° F. (100,000,000° C), a new type of nuclear fusion will occur, in which helium atoms combine to produce carbon.

When the sun's helium "ashes" finally ignite, a tremendous explosion will probably blow away much of the sun's outer layers. In time, fierce solar winds will blow all but the core itself into space. All that will be left of our star will be a hot, dense remnant called a white dwarf, about the same size as Earth and perhaps one-thousandth as bright as the sun today.

During billions of years as a white dwarf—astronomers can't say how long with any certainty—the sun will continually cool down, and it will eventually become a cold, dark ember of the glorious beacon it is today.

THE SOLAR SYSTEM

On our tour of the realm of the sun, we will explore the various worlds that make up the solar system. Giant worlds made of gas and ice, rocky worlds scorched by the sun, and cratered worlds with barely a smooth patch of ground all exist within the same system of planets and moons. And, of course, so does Earth—a planet unlike any other.

It seems almost unbelievable that such varied worlds can exist in the same solar system. Even planetary scientists and astronomers are often amazed by new discoveries, such as giant volcanoes and canyons on Mars, active volcanoes on Jupiter's moon Io, and a huge storm system on Neptune.

The story of how these worlds formed is hidden from us. Various theories have been suggested to describe the formation of the solar system. Despite the recent intensive exploration of the sun and planets, both from the Earth and with spacecraft, there is still no universal agreement on any of these theories. The following outline of the formation of our solar system is a brief description of one of the leading theories. It will probably be modified as our knowledge of the solar system continues to grow.

Our solar system probably formed from a vast cloud of gas and dust (upper left). This cloud collapsed and began spinning around a central bulge (far left). The central bulge became the sun, and swirling eddies of gas and dust continued to orbit around it (left). These whirlpools of gas and dust eventually formed into the planets (below).

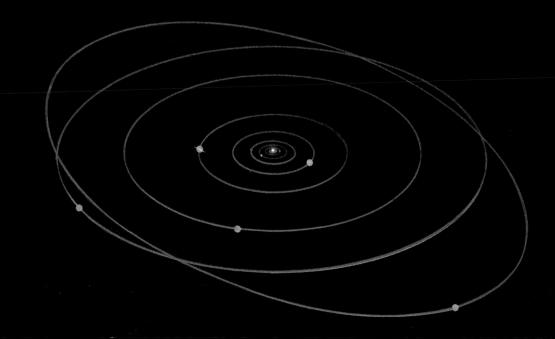

A BEGINNING IN DARKNESS

About 4.6 billion years ago, a vast, slowly rotating cloud of gas and dust began to come together. It was not unlike many interstellar clouds we see today, including the Orion nebula.

No one is sure what caused the cloud to collapse on itself. One plausible theory says that the explosion of a nearby star sent shock waves across interstellar space and triggered the collapse.

As it collapsed, the rotating cloud began to spin faster and faster, the way figure-skaters spin faster when they draw their arms in. The cloud continued to contract and whirl until it formed a spinning disk the size of our solar system. Scientists call this disk of gas and dust the solar nebula. At the center of the nebula was a relatively dense bulge that would eventually form into the sun.

The collapse of the interstellar cloud created a great deal of heat, and soon after it formed, the solar nebula was quite hot. In its inner regions, interstellar dust grains were probably vaporized by the intense heat. The center of the nebula got hottest, and when temperatures reached some 18,000,000° F. (10,000,000° C), thermonuclear reactions were triggered in the central bulge, and the sun was born.

FIRE AND ICE

Meanwhile, the nebula itself was condensing and forming various substances. Because temperatures varied across the nebula, different substances formed at different distances from the sun. In the outermost regions, far from the infant sun, volatile compounds such as water, ammonia, and methane, which vaporize between 212° F. and 572° F. (100° C and 300° C), condensed to form ice crystals. Closer in, temperatures of about 1,800° F. (1,000° C) were too high for ice to form but low enough for grains of silicate dust to condense. Still closer to the newborn sun, iron, aluminum, and titanium, which vaporize between 4,440° F. and 5,916° F. (2,467° C and 3,287° C), condensed into grains of metal.

We see the effects of this temperature range today. In the inner solar system, we find dense, rocky bodies. Earth, Venus, and Mercury are the richest in metal, similar in density, and presumably started out with similar compositions. But as we shall see, torrid Venus turned out very differently from Earth. Moving farther out, we find Mars, a rocky planet less dense than Earth. Beyond Mars are the asteroids—leftover chunks of rock and metal. And in the outer solar system, gas and ice dominate. Jupiter, Saturn, Uranus, and Neptune are rich in the lightest elements hydrogen and helium, while their satellites are rich in ice.

THE BIRTH OF THE PLANETS

As the nebula condensed, small bodies called planetesimals took form. Within a few thousand years, there were millions of these bodies circling the infant sun, each a few miles across. Small dust particles stuck to these circling bodies, and they got larger. These planetesimals collided and stuck together, and within a few million years grew to be perhaps 60 miles (100 kilometers) across. Their gravity attracted more debris, and the embryonic planets slowly grew to their present size.

During this period, the planets were subjected to an unending rain of debris. According to researchers, this bombardment was so violent that the surfaces of the solid planets may have become completely melted. After their surfaces cooled and solidified again, the newly formed planets quickly swept up the last wandering planetesimals, producing the crater-covered surfaces seen on Mercury, the moon, Mars, and the satellites of Jupiter and Saturn.

The record of this violent bombardment has been erased on Earth by geologic activity. On the moon, Mercury, and Mars, only part of the early record was erased before these planets grew cold and their geologic activity ceased.

COMETS: GIVERS OF LIFE

At the fringes of the solar nebula, where the temperature was cold, condensation formed the fluffy lumps we call comets. This material was unaltered by the subsequent history of the system, and these frozen wanderers may hold the key to deciphering the earliest events in the solar system's formation. They may allow us to reach back 5 billion years or more in time in order to find out what things were like in the early solar nebula. They may also have been responsible for life itself. Even as Earth and the other solid planets were forming, comets had formed at the solar system's edge. The gravity of Jupiter and other massive planets probably attracted many comets into the inner solar system. Earth and the other solid planets swept up these comets, incorporating their water and other elements.

The moon, Mars, Venus, and Mercury appear to have lost most or all of this supply of wa-

ter and other elements. But Earth held onto these elements in the form of an atmosphere and oceans, allowing life to arise.

OTHER SUNS, OTHER PLANETS

The interstellar cloud that spawned the solar nebula may have contained a thousand times the total mass of our solar system. It may have fragmented to produce a number of stars and solar sytems. Stars are being born from such clouds even today. There are probably planets around some of them, because the formation of planets seems to be a natural occurrence. Nothing special is required. Astronomers finally have several high-quality observations showing the presence of planets around some nearby stars. And the laws of probability indicate that some of these may harbor life.

The search for life on extrasolar planets is one of astronomy's most exciting challenges. While no confirmed extrasolar life forms have been found, astronomers continue to monitor the skies searching for life on other planets. Some astronomers use giant radio telescopes to listen for signals that might be sent from intelligent beings from extrasolar civilizations.

Still, there are plenty of wonderful and fascinating worlds within our own solar system that should hold astonomers' interest for years to come. It is to these planets, moons, asteroids, and comets that we now turn our attention.

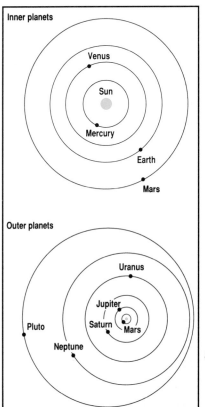

The orbits of the nine planets are shown in this double diagram. The top diagram plots the orbits of the four inner, terrestrial planets. The lower one depicts the orbits of the four gaseous planets and Pluto. Because Pluto's orbit is very elliptical, the planet passes within the more circular orbit of Neptune whenever Pluto is at perihelion (point closest to the sun). Pluto passed within Neptune's orbit in 1979 and will remain there until 1999. Pluto and Neptune never collide because their orbits are inclined differently and never truly intersect.

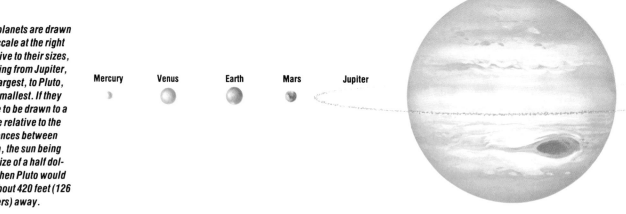

The planets are drawn to a scale at the right relative to their sizes, ranging from Jupiter, the largest, to Pluto, the smallest. If they were to be drawn to a scale relative to the distances between them, the sun being the size of a half dollar, then Pluto would be about 420 feet (126 meters) away.

Mercury Venus Earth Mars Jupiter

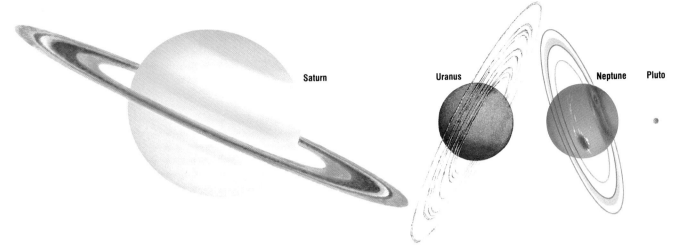

Saturn Uranus Neptune Pluto

This is an artist's rendition of how our solar system might look if each of the planets were visible from Pluto (lower right). In reality, the four inner planets would probably not be visible, the distances between the planets would be much greater, and the sun, approximately 3.7 billion miles (5.9 billion kilometers) away, would be barely larger than other neighboring stars.

39

EARTH AND THE MOON

Earth is enclosed in a bluish envelope of air, called the atmosphere, which extends as far as 1,000 miles (1,600 kilometers) above the surface.

EARTH

If you were approaching the sun from another star, you probably wouldn't notice the small planet called Earth. Huge Jupiter might catch your eye first. Or perhaps Saturn, with its beautiful system of rings, would attract your attention. But if, in your wanderings, you visited the third world out from the sun, you would

The workings of both nature and humans are evident in landforms on Earth. This satellite photograph of Medicine Hat in Alberta, Canada, shows the carving effects of rivers and streams, the quiltwork appearance of cultivated fields, and the crisscrossing of city and country streets and roads.

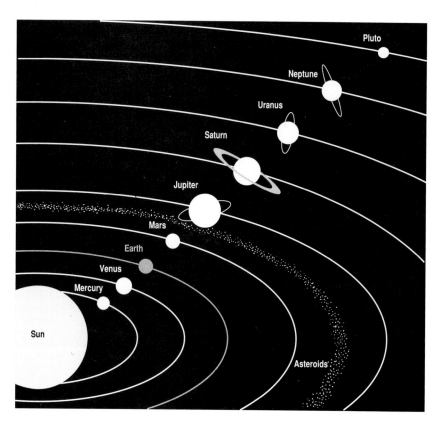

find that it is unlike any other in the solar system. The most important difference can be said in a single word: life. No other planet is known to harbor living things.

42　This is no coincidence. Conditions on Earth are just right for life as we know it. Our atmosphere contains the right mixture of gases. Temperatures are moderate over most of the planet. And that precious substance called water is plentiful in liquid form.

Earth itself is alive with geologic activity. Wind and water are in constant motion, changing the face of the land. The land itself is stirring, with ranges of mountains being pushed up and volcanoes pouring forth molten rock to make new land. From its beginning, Earth has been an active planet.

What was that beginning? The Earth's crust has changed so much over the eons that all traces of our planet's birth and infancy have been erased. The oldest rocks found on Earth are 4.3 billion years old, yet geologists know that our planet, and the rest of the solar system, formed some 800 million years earlier. For scientists, piecing together that missing history and the story of Earth's origin requires painstaking detective work.

How Earth Began

According to current ideas, Earth, like the sun and the other planets, condensed out of a cloud of gas and dust that slowly began to contract roughly 4.6 billion years ago. As the cloud shrank, some of the dust grains, coated with a fluffy layer of frozen cloud material, gently collided and stuck together. More and more grains collided and accumulated. Soon there were a number of small, solid lumps called planetesimals orbiting amid the gas and dust, and these collided with each other to form still bigger fragments.

Within a few thousand years, enough planetesimals collided to form an embryonic planet several miles across. A hundred million years or so later, our world had grown to roughly its present size. Its interior was a fairly uniform mixture of dust and frozen gases. But it did not stay that way.

Each particle that collided with the growing Earth added some energy to the planet. More heat came from the decay of radioactive elements (like isotopes of potassium, thorium, and uranium) in its interior. Eventually, temperatures climbed so high that Earth's interior melted completely through.

Imagine this molten planet that existed some 4.5 billion years ago. Within the liquid, elements of different density separated: heavier components, such as iron and nickel, sank to the middle of Earth, while lighter ones such as silicon, aluminum, and magnesium, stayed closer to the surface. Lightest of all were gases such as nitrogen, hydrogen, and oxygen.

These lightest components rose to the uppermost layers of the molten interior. Later on, these gases escaped to the surface through volcanoes and became part of the atmosphere. The water vapor condensed to form Earth's oceans. About 3.7 billion years ago, photosynthesis by early organisms began changing carbon dioxide in the atmosphere into oxygen. Most of the rest became incorporated in limestones and other carbonate rocks.

The Structure of Earth

What lies within Earth today? Geologists have explored the planet's interior by studying volcanoes and volcanic rocks, by examining the composition of meteorites, and by investigating earthquakes. When an earthquake occurs, it sends seismic waves through Earth. These waves change in frequency and intensity as they pass through the different layers of the planet. From their analyses of earthquakes, geologists believe Earth is comprised of three main components—the core, the mantle, and the crust.

The core, with a radius of approximately 2,200 miles (3,500

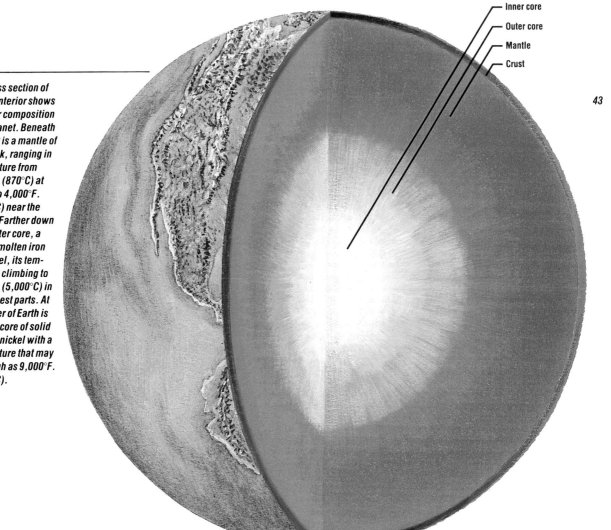

This cross section of Earth's interior shows the inner composition of the planet. Beneath the crust is a mantle of solid rock, ranging in temperature from 1,600°F. (870°C) at the top to 4,000°F. (2,200°C) near the bottom. Farther down is the outer core, a layer of molten iron and nickel, its temperature climbing to 9,000°F. (5,000°C) in the deepest parts. At the center of Earth is an inner core of solid iron and nickel with a temperature that may be as high as 9,000°F. (5,000°C).

Inner core

Outer core

Mantle

Crust

43

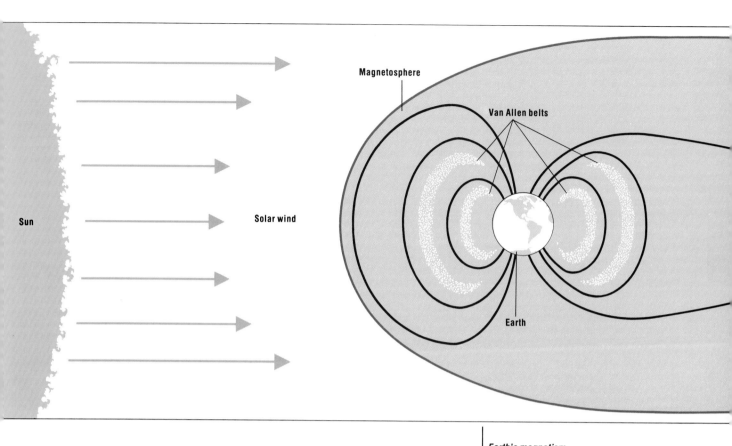

Magnetosphere

Van Allen belts

Sun

Solar wind

Earth

kilometers), is probably a mixture of iron and nickel. The center of Earth is still very hot—so hot that the outer portion of the core is actually metallic liquid. The motion of this fluid as Earth rotates may be the cause of the planet's magnetic field. The inner part of the core is also extremely hot, but the immense weight of the rest of Earth squeezes it into a solid ball.

The mantle, which is approximately 1,800 miles (2,900 kilometers) thick, forms most of the outer half of Earth. It is made of dense rocks rich in magnesium, iron, and silicon. The outermost portion of the mantle is somewhat soft, and it actually circulates in huge, slow-moving currents beneath the crust. These movements, known as convection currents, are the means by which the intense heat of Earth's interior is carried to the surface.

The crust is the rigid, outermost layer of Earth. It is broken up into a series of plates that ac-tually move across the face of the globe, propelled by the convection currents in the upper mantle. The continents and ocean floors, which ride atop these plates, are in constant motion, but you would never no-tice—they move about as fast as your fingernails grow. Millions of years ago, the continents oc-cupied very different positions than they do today.

The Ever-Changing Crust

Two types of crust cover Earth: the oceanic crust and the conti-nental crust. The oceanic crust, which forms the ocean floors, is mainly made of medium-dense igneous rocks—basalt, gabbro, and peridotite. Oceanic crust is about 5 miles (8 kilometers) thick on the average.

At vast fissures in the ocean floor, called mid-ocean ridges, molten rock flows out from the mantle to form new crust. Over millions of years, the ocean floor

Earth's magnetism operates in a region called the magneto-sphere. Radiation from the sun acts like a wind that blows the magnetosphere into a long tail. The magnet-osphere includes the Van Allen belts, which consist of large num-bers of charged elec-trons and protons trapped by the mag-netism. (The distance between the sun and Earth is not to scale.)

crawls away from this ridge like a conveyor belt, until it meets one of many great trenches called subduction zones. At the subduction zones, older crust ac-tually descends into the mantle, where it melts and is recycled for new crust. Because oceanic crust is constantly being recy-cled, the oldest rocks on the ocean floor are only about 200 million years old—youthful

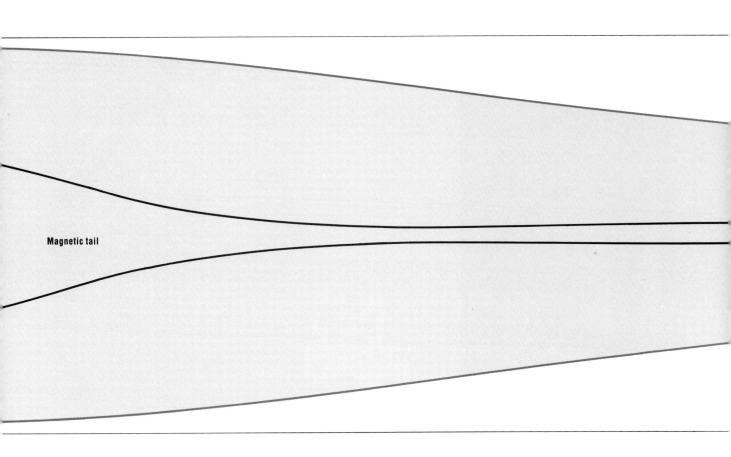

Magnetic tail

compared to the 4.6-billion-year-old Earth.

In addition to the rocks that lie under the ocean, the continental crust has another layer of lighter, silicon-rich granite, plus further layers of a wide variety of fairly lightweight rocks. It is also quite a bit thicker than the oceanic crust, with an average of 25 miles (40 kilometers). Under some mountain ranges, the crust is as much as 55 miles (90 kilometers) thick.

Because the upper layer of the continental crust is of very low density, it is too buoyant to be dragged down with the descending crustal plate underneath. So, the continents ride atop the crustal plates, grinding past each other in some places (causing earthquakes in the process) and colliding to form great mountain belts (such as the Himalayas) in others. Some recycling may take place, but it is not nearly as dramatic as that of the oceanic crust.

Most of Earth's large-scale features, including mountain ranges, rifts in the crust, and chains of volcanic islands, can be explained by the motions of the crustal plates, a process that geologists call plate tectonics.

The planet we call home is special, not just because it is our home, but because it is an active, evolving world. As we shall see, Earth has outlived most of the other solid planets, whose internal fires apparently died out long ago, leaving cold, dead worlds. Earth is, by comparison, the jewel of the solar system.

45

This diagram shows how hot rock rising beneath the Mid-Atlantic Ridge adds material to the plates of South America and Africa. Where two plates meet, the old material is either pulled down into the earth's mantle or pushed up to form mountains. The Peru-Chile trench and the Andes Mountains were formed by the South American plate and a Pacific Ocean plate.

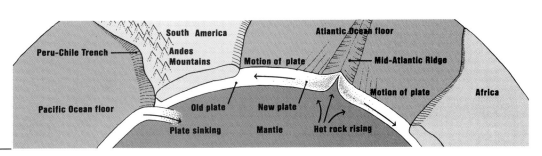

Pythagoras W. Bond Strabo

Sea of Cold
(Mare Frigoris)

Endymion

Bay of Dew

Plato Aristoteles

Alps Alpine Valley Hercules Atlas

Jura Mountains Mount Eudoxus Franklin
Bay of Straight Pico
Rainbows Range Messala

Mairan Aristillus Lake of Dreams Geminus

Sea of Rains
(Mare Imbrium) Caucasus Posidonius Taurus
Mountains Cleomedes
Archimedes Autolycus
Timocharis Marsh of Apollo 15 landing Macrobius
Decay Sea of Serenity
Struve Ocean of Storms Aristarchus (Mare Serenitatis) Sea of Crises
Apollo 17 landing (Mare Crisium)
Apennines Haemus Mountains
Carpathians Plinius Marsh of Sleep

Eratosthenes Sea of Vapors Marginal Sea
Cardanus (Oceanus Procellarum) (Mare Manilius
Copernicus Seething Vaporum)
Bay Sea of Tranquility Foaming Sea
Kepler Julius Caesar (Mare Tranquillitatis)
Riccioli Reinhold Agrippa Maskelyne
Central Schmidt Apollo 11 landing
Lansberg Bay Sea of Fertility
Hipparchus Delambre (Mare Fecunditatis)
Riccioli Apollo 12 landing Gutenberg Smyth's Sea
Grimaldi Apollo 14 landing Apollo 16 Theophilus Langrenus
Letronne landing Albategnius Cyrillus Goclenius
Fra Mauro Ptolemaeus Pyrenees
Gassendi Abulfeda Sea of Nectar Colombo
Alphonsus (Mare Nectaris) Vendelinus
Mersenius Arzachel Catharina Santbech
Sea of Moisture Sea of Clouds Petavius
(Mare Humorum) (Mare Nubium) Straight Wall Fracastorius
Altai Scarp Humboldt
Mercator Pitatus Purbach Furnerius
Campanus Regiomontanus Piccolomini
Deslandres Walter
Wilhelm Orontius Stöffer Maurolycus
Schickard Janssen
Longomontanus Tycho Pitiscus
Maginus
Schiller Cuvier Vlacq
Scheiner Clavius
Curtius Manzinus
Blancanus

46

THE MOON

Our nearest neighbor in space has awed humanity from the beginning of time. Through the ages, the moon has been a god to some, a demon to others, and a mystery to all. Whether a half-circle bathed in the blue of an afternoon sky, a slender crescent hovering above the treetops at dusk, or a round, midnight beacon shining at full strength overhead, the moon dominates our skies. Today we view the moon not as a spirit, but as another world. What do we know of this alien body?

To the naked eye, the moon is a collection of light and dark patches. To many, the pattern resembles a face—the face of the man in the moon. To some, the moon has appeared to be a smooth, polished crystal sphere.

When Galileo turned his new telescope on the moon in 1609, he saw surprising new details. The surface was not smooth, as it looks from Earth, but covered with holes, which we call craters. Galileo saw that the dark

The nearside of the moon is the only side we ever see from Earth. This is because the moon turns once on its axis in the same time that it circles Earth, thus always keeping the same side facing Earth.

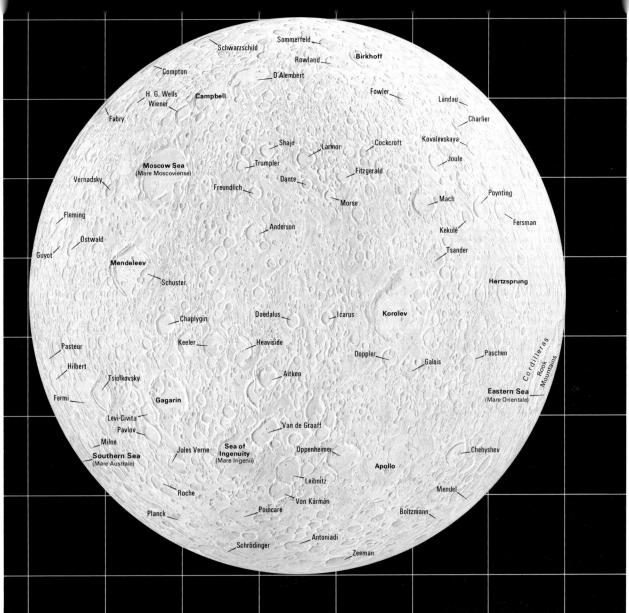

The far side of the moon was seen for the first time in 1959 when the Soviet Union's Luna 3 sent pictures back to Earth.

patches were smoother than the brighter, heavily cratered areas. Believing these smooth areas to be seas, he named them *maria* (singular, *mare*), the Latin word for seas.

In the centuries following Galileo, the moon was a new world to explore for astronomers armed with ever larger telescopes. By the middle of this century, they had seen that our satellite is a very different place than Earth. The moon has no atmosphere, no life. The maria turned out to be seas, not of water, but of

hardened lava. And thousands of craters cover the barren, unchanging moonscape.

For many years, scientists lost interest in the moon. But by the middle of this century, astronomers realized that the moon is a museum piece of priceless value. All the processes that have erased the earliest geologic record of Earth—erosion by wind and water, distortion of the crust by folding, recycling of the crust by plate tectonics—are absent on the moon. On its crater-pocked surface is written the early history of our solar system, and the tale is a violent one.

The Moon's Violent Past

After the sun and planets formed, the solar system was filled with leftover debris. Each planet was a wandering target for chunks of rock called meteoroids. Meteoroids slammed into the moon at such high speeds that they blasted great holes, called impact craters, in the crust. Craters are the most common landform on the moon and, for that matter, in the entire solar system.

For nearly a billion years, meteoroids of all sizes rained onto the lunar surface, creating thousands of craters. Then, about 4 billion years ago, the bombardment died down. Today, small meteorite particles still strike the moon's surface. However, impacts by large bodies are a rare occurrence. But a look through a small telescope or a pair of binoculars shows the scars of the earlier, more violent era.

One of the more prominent lunar impact craters is Copernicus, which was formed roughly 900 million years ago. An explosion with as much energy as 20 trillion tons of TNT created this crater, 56 miles (90 kilometers) wide, in a matter of seconds. And Copernicus is only a medium-sized crater.

The largest craters, called impact basins, are hundreds of miles (kilometers) across. The impacts that formed these must have been truly colossal. After the basins formed, lava rose through cracks in the moon's crust and flooded their centers, creating the maria. Subsequent smaller impacts formed craters on these smooth lava plains.

The distribution of lava plains on the moon is puzzling to scientists. Almost all are on the side that always faces Earth (called the nearside). Only a few

48

This composite photograph of the moon shows the wide variety in size and shape of lunar craters.

are on the far side. This uneven distribution has not been completely explained. It may be that the nearside crust is thinner than that of the far side. If so, lavas from the interior may have broken through to the surface more easily on the nearside than the far side.

The Message in Moon Rocks

In 1969, astronauts landed on the moon and brought back treasures of incredible value—lunar rocks. Now scientists could study the moon directly instead of just in photographs. The lunar samples showed that the moon's surface is very, very old. The youngest lunar rocks yet analyzed—volcanic rocks called basalts from the maria—are 3.2 billion years old.

The heavily cratered highlands contain some of the oldest rocks on the moon—formed about 4.4 billion years ago, just 200 million years later than the solar system itself. By comparison, the oldest rocks yet found on Earth are only 3.8 billion years old. Geologists believe the moon became so hot during its formation that its upper layers became molten. Lighter minerals floated to the surface and formed a crust made of a rock called anorthosite. The highlands are the remains of that early lunar crust.

Over billions of years, impacts have changed the original crust. Samples of the old anorthosites are hard to find. Another type of rock is more common—breccia,

Hipparchus (180 B.C.?–125 B.C.?) was an ancient Greek astronomer who calculated the length of the year to within six minutes. His observations of Earth, the moon, and the sun greatly influenced later astronomers, including Ptolemy.

49

which is a collection of dust and rock fragments welded together by the heat and pressure of a meteoroid impact. Many of the rocks in and around impact craters are breccias.

Where Did the Moon Come From?

The most basic question about the moon—how it formed—is still unanswered. Some scientists believe the moon broke away from Earth while our planet was still molten. But it is difficult to explain how the material that formed the moon was able to escape from Earth in the first place.

Other scientists have proposed that the moon and Earth formed separately from the same cloud

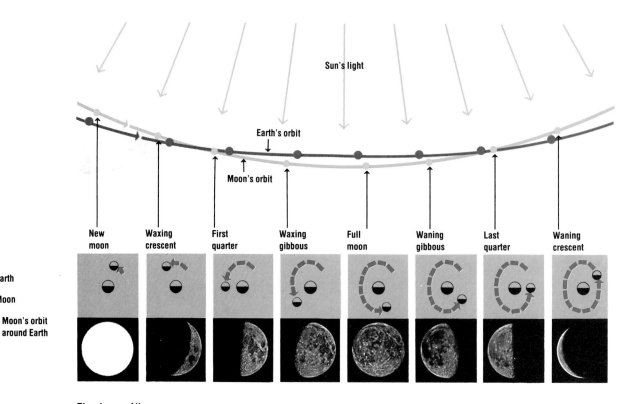

Sun's light

Earth's orbit

Moon's orbit

| New moon | Waxing crescent | First quarter | Waxing gibbous | Full moon | Waning gibbous | Last quarter | Waning crescent |

● Earth

○ Moon

↤ Moon's orbit around Earth

The phases of the moon are caused by the moon's orbit around Earth as Earth and the moon travel around the sun. Half of the moon is always in sunlight, but varying amounts of the lighted side are visible from Earth. As the moon and Earth move along their orbits, more of the sunlit part is seen, until it shines as a full moon. Then less and less of the sunlit part is seen until the dark new moon returns. (The Earth and the moon are not drawn to scale.)

of gas and dust. The problem here is that the moon's composition is somewhat different from that of Earth, a fact that is difficult to explain if both bodies came from the same source material.

A third theory holds that the moon originally formed far from Earth but later wandered close enough to be captured by our planet's gravity. The problem here lies in explaining how such a capture occurred and where in the solar system it formed to have its present composition.

Recently, a theory proposed in the mid-1970's has gained the greatest acceptance from astronomers. In this view, the interior of the young Earth melted, forming an iron-rich core and an iron-poor outer half (the mantle). Sometime later, one of the large meteoroids, perhaps thousands of miles across, struck Earth and blasted a cloud of debris from our planet's upper layers into space. This cloud ultimately condensed to form an iron-poor moon with some of the chemical properties of Earth's mantle.

Compared to Earth, the moon is a primitive, unevolved body. It was once believed that the moon was of the same composition throughout. Since the lunar landings, we now know that the moon, like the planets, is composed of layers. There is a thin crust, perhaps 35 to 60 miles (56 to 96 kilometers) thick. There is a mantle, homogeneous in composition, that extends to approximately 185 to 310 miles (296 to 496 kilometers) toward the center. The core, which is small, may be iron-rich, but nonmolten.

Unlike Earth, the moon lacked the necessary bulk to retain its heat for long. After the last of the mare basalts oozed onto the surface about 3 billion years ago, the moon was essentially a dead world, almost exactly as we see it today. Fortunately, it has much to tell us of its past, as well as our own.

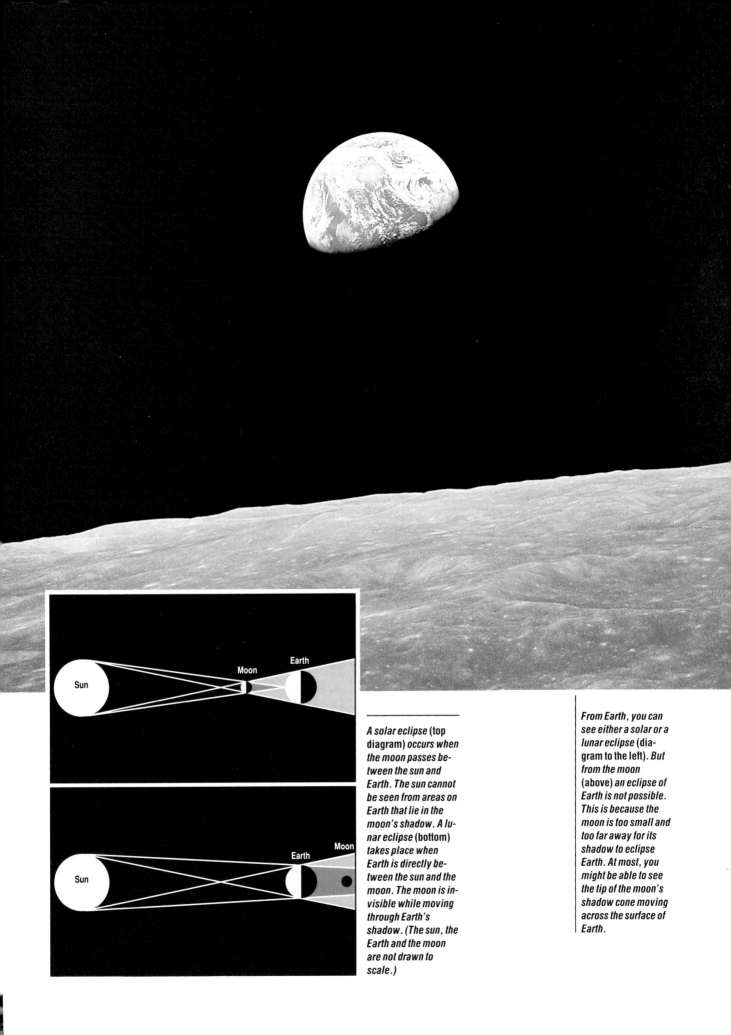

A solar eclipse (top diagram) *occurs when the moon passes between the sun and Earth. The sun cannot be seen from areas on Earth that lie in the moon's shadow. A lunar eclipse (bottom) takes place when Earth is directly between the sun and the moon. The moon is invisible while moving through Earth's shadow. (The sun, the Earth and the moon are not drawn to scale.)*

From Earth, you can see either a solar or a lunar eclipse (diagram to the left). But from the moon (above) an eclipse of Earth is not possible. This is because the moon is too small and too far away for its shadow to eclipse Earth. At most, you might be able to see the tip of the moon's shadow cone moving across the surface of Earth.

Mercury was photographed in detail for the first time by Mariner 10 in 1974. Eighteen photographs were computer-enhanced and put together to make this photomosaic of the planet's cratered surface.

MERCURY

Imagine discovering a new planet overnight. That is essentially what happened to scientists in 1974, when the United States spacecraft *Mariner 10* flew past Mercury. Until then, this planet closest to the sun—roughly two-fifths the size of Earth—had been seen as no more than a tiny, almost featureless blur in even the largest telescopes. From Earth, Mercury can often be seen at dawn or dusk. However, it is sometimes difficult to find because it never strays far from the glare of the sun.

Mercury is a world ruled by the sun. If you could stand on its surface, the sun would be a fiery disk two and one-half times bigger than seen from Earth. The sky would be black, for Mercury has only the barest wisp of an atmosphere. No blanket of air shields Mercury's surface from the sun's intense heat and deadly radiation. During the day, temperatures soar as high as 801° F. (427° C). Human explorers would find it extremely difficult to survive, even in space suits. At night, temperatures plummet to −279° F. (−173° C). Mercury is clearly a world of extremes.

All this was known before *Mariner 10* flew past Mercury. Astronomers also knew that the planet's density is unusually high—more than five times that of water—which implies that Mercury has a large, iron-rich core. And by bouncing radar waves off the planet, they had determined that Mercury rotates once every 59 earth-days. But what was Mercury's surface like? Did it have craters such as those on the moon? How much had Mercury evolved since its formation? Until *Mariner 10*'s arrival, little could be said to answer these questions.

Mariner 10 ended Mercury's reign as an unknown world. In its three fly-bys, the robot explorer sent back detailed pictures of almost half the planet. It confirmed the high surface temperatures and found a weak magnetic field. *Mariner 10* data also confirmed that Mercury's core is composed of metallic iron, which comprises 70 percent of the planet's mass.

The pictures show a barren world that, like the moon, is covered with impact craters. The fact that there are so many craters means that much of Mercury's surface is very old, probably close to the age of the solar system itself.

Unlike the moon, the portion of Mercury photographed by *Mariner 10* has few large impact basins, and almost no broad lava seas. One large, spectacular basin, called Caloris, was found. It is a multiringed crater some 800 miles (1,280 kilometers) in diameter. Within Caloris are lava plains that resemble the lunar maria. These lavas must have flooded the basin sometime after it was blasted out of Mercury's crust.

Many of Mercury's craters have smooth areas between them that also look like volcanic plains. Some may indeed be lava, or they may be blankets of debris ejected from nearby impact craters. The origin of these smooth intercrater plains is still unknown.

Mariner 10 found surprising, fascinating features—a number of clifflike ridges stretching thousands of miles across Mercury. Some of these scarps cut right through impact craters. Geologists believe the scarps were created early in Mercury's history, after the planet's crust had formed but before the interior had cooled off.

Mercury appears to be a primitive world. Like Earth, Mercury probably became hot enough to melt soon after it formed, and its material separated into layers. But geologists do not think Mercury evolved much further before its interior cooled down. The planet appears to have died geologically early in its history.

Mercury at a glance

Mercury, shown in orange in the diagram, is the closest planet to the sun.

Distance from sun: *Shortest*—28.6 million miles (46 million kilometers); *Greatest*—43.4 million miles (69.8 million kilometers); *Mean*—36 million miles (57.9 million kilometers).

Distance from earth: *Shortest*—57 million miles (91.7 million kilometers); *Greatest*—136 million miles (218.9 million kilometers).

Diameter: 3,031 miles (4,878 kilometers).

Length of year: 88 earth-days.

Rotation period: 59 earth-days.

Temperature: −279° to 801° F. (−173° to 427° C).

Atmosphere: Sodium, helium, hydrogen, oxygen.

Number of satellites: None.

(Planets and orbital distances are not to scale.)

54 *Venus, as seen from Earth, is brighter than any other planet or star. At certain times of the year, Venus is the first planet or star that can be seen in the western sky in the evening. At other times, it is the last planet or star that can be seen in the eastern sky in the morning.*

VENUS

Venus has long been a world shrouded in mystery. Brilliant yellowish-white clouds of sulphuric acid completely surround the planet, hiding its surface from our view. For decades, astronomers knew nothing about Venus except its most basic attributes. In size and density, Venus is almost a twin of Earth. Was it possible that beneath the clouds, another earth existed? Astronomers knew that Venus, being roughly 30 percent closer to the sun than Earth, would be a good deal warmer. Perhaps, some speculated, Venus was covered with tropical forests, much like the prehistoric Earth. They couldn't have been more wrong.

When the American space probe *Mariner 2* flew by Venus in 1962, it sent back verification of data collected earlier by radio astronomers: Venus' carbon dioxide atmosphere was far too hot to support life. More details on the mysterious world were slow in coming. Finally, in the 1970's, the Soviet Union's *Venera* probes landed on Venus and sent back the first reports from its surface. They found truly hellish conditions: a bare, rock-strewn landscape lay beneath crushing atmospheric pressure 90 times that on Earth's surface—equivalent to pressures found in the depths of Earth's oceans—at a scorching temperature of 864° F. (462° C). In fact, Venus is even hotter than its sunward neighbor, Mercury.

THE GREENHOUSE PLANET

Why is Venus so hot? The answer is that the planet's massive carbon dioxide atmosphere acts like a greenhouse. The glass walls of a greenhouse allow sunlight to pass through, warming its contents during the day. Objects inside reemit the energy as heat, but the glass does not transmit this infrared (heat) radiation. As a result, greenhouses stay warm even when their surroundings are cool. Because carbon dioxide readily absorbs infrared radiation, Venus' greenhouse is a very efficient one, maintaining the planet's extremely high temperatures. The dense blanket of gases and high-speed winds also carry heat to the planet's night side, which is almost as hot as the day side.

Venus could not be any less Earthlike. Yet, scientists find it puzzling that two planets as similar in size and mass should have turned out so differently. One important fact seems to be Venus' lack of liquid water, which strongly influences climate. Venus may have started

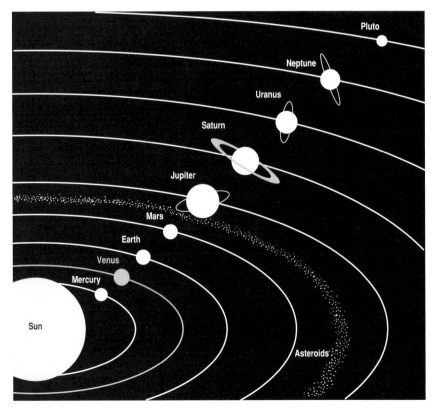

Venus at a glance

Venus, shown in orange in the diagram, is the second closest planet to the sun.

Distance from sun: *Shortest*—66.8 million miles (107.5 million kilometers); *Greatest*—67.7 million miles (108.9 million kilometers); *Mean*—67.2 million miles (108.2 million kilometers).

Distance from earth: *Shortest*—25 million miles (40.2 million kilometers); *Greatest*—160 million miles (257 million kilometers).

Diameter: 7,521 miles (12,104 kilometers).

Length of year: 225 earth-days.

Rotation period: 243 earth-days.

Temperature: 864° F. (462° C).

Atmosphere: Carbon dioxide, nitrogen, water vapor, argon, carbon monoxide, neon, sulfur dioxide.

Number of satellites: None.

(Planets and orbital distances are not to scale.)

out with nearly as much water in its bulk as Earth did. But it may be that, due to the planet's nearness to the sun, virtually all of Venus' water was rapidly boiled off into the atmosphere, where it was then broken down by sunlight and irretrievably lost. We can be very thankful for the fact that Earth held onto its supply of water; if it hadn't, a carbon-dioxide atmosphere and Venuslike conditions would surely have developed. But as it turned out, most of Earth's carbon dioxide is contained in limestone rocks or is dissolved in the oceans.

Today, the atmosphere on Venus is nearly all carbon dioxide, with small amounts of nitrogen and only minute amounts of sulfur dioxide, water vapor, carbon monoxide, and other gases. There are even traces of hydrochloric acid.

Were you able to stand on Venus' surface, you would not see the sun. Instead, the sky would be an unbroken, brightly lit cloudscape. Because the thick atmosphere absorbs all but the redder wavelengths of sunlight, everything would look orange, as if you were wearing tinted sunglasses. Even in a well-insulated, pressure-resistant spacecraft, you probably would not want to stay long. Venus is not a very inviting world.

Venus' atmosphere receives so much energy from the sun and the greenhouse effect that it races around the planet once every four Earth-days. For comparison, the planet itself takes 243 Earth-days to spin once on its axis. In addition, Venus is the only planet that does not rotate in the same direction in which it travels around the sun. This odd retrograde (opposite) rotation has not yet been conclusively explained.

In visible light, the clouds are featureless. But seen in the ultraviolet, they display broad, Y-shaped patterns caused by the atmosphere's swift motion. Though study of the Venusian atmosphere is still in its beginnings, scientists hope to find clues to the behavior of our own atmosphere by studying that of Venus.

BENEATH THE CLOUDS

What lies beneath the gleaming white clouds? There is an answer, for scientists have finally lifted the Venusian veil—with radar. American and Soviet spacecraft orbiting Venus have used radar waves to penetrate the clouds and create a picture of the planet's surface. Ground-based radar systems have also contributed to this mapping process.

Interestingly, Venus has relatively few impact craters. On a large scale, it looks surprisingly like Earth: low plains similar to Earth's ocean floors cover much of the globe. These are broken here and there by vast plateaus that resemble continents.

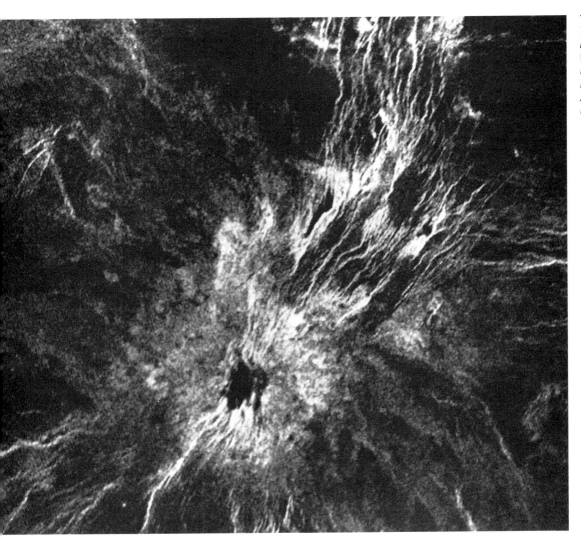

57

In more detailed views, Venus displays a host of complex and varied landforms, many unlike anything seen on Earth or other worlds. The continents contain jumbled, fractured mountain ranges, in some areas resembling the crinkled hide of an elephant. Some of Venus' mountains tower high into the broiling atmosphere, higher than Earth's Mount Everest. Enormous rifts in the Venusian crust have also been detected.

The radar pictures also show that Venus has been volcanically active. Lava flows are widespread, and huge features resembling shield volcanoes have been spotted. There are also broad, oval-shaped features that appear to be upwellings of lava at shallow depths in the Venusian crust.

Hardly anything is known of Venus' internal structure. Based on its calculated density and size, scientists believe the planet probably has a solid core composed largely of nickel and iron. Any liquid core must be quite small, since Venus possesses no magnetic field. Most of the planet's bulk is likely contained in a thick mantle, topped by a crust thought to be thicker than Earth's.

In 1982, two more *Venera* probes provided the first detailed chemical analyses of the surface. Each of the craft drilled into the surface, extracted a small sample, and analyzed it in a sophisticated on-board laboratory. The results show that the places where the *Veneras* landed are covered with rocks similar in composition to volcanic rocks called basalts that are found on Earth.

A PRIMITIVE EARTH?

As yet, scientists have no solid evidence that Venus' crust may be broken up into moving plates like Earth's. Like the moon and Mercury, it may be another one-plate planet, but it has certainly been geologically active. It appears that the Venusian surface has been flooded by volcanic lavas, fractured, and folded. Fewer impact craters than expected have been found, suggesting either that meteors burn up before they can strike the surface, or that much of the planet's terrain formed less than a billion years ago. Venus' earlier history has almost surely been erased. And

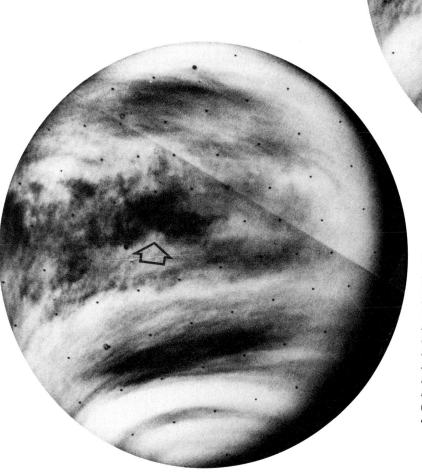

These three photographs of Venus were taken at seven-hour intervals by Mariner 10 in 1974, showing how rapidly its thick cloud cover moves across the planet's surface. The feature indicated by the arrow is about 600 miles (1,000 kilometers) across.

it is quite possible that Venus is still active today. Some scientists believe that geologically Venus is just getting started. In that case, it may be much like the early Earth was.

Venus has not yielded easily to the curiosity of human beings. Its opaque clouds force us to use unusual methods for glimpses of its surface. Its hot, corrosive, crushing atmosphere destroys our space probes even as they radio their findings to Earth. Undaunted, scientists continue to dispatch new spacecraft to Venus. In 1985, two more Soviet *Venera* probes landed on the planet's nightside to report on conditions and analyze the surface. They were accompanied by French-made balloons that were released into the Venusian winds, sending back data on the upper atmosphere. And the United States has also sent a space probe to Venus. On May 4, 1989, space shuttle *Atlantis* launched spacecraft *Magellan* which will reach Venus during 1990.

Venus goes through phases similar to those of the moon. However, in addition to changes in brightness and shape, Venus seems to change in size also. This is due to its changing distance from Earth. Venus appears largest when on the same side of the sun as Earth, at which time only a thin, sunlit crescent-area can be seen. Venus is smallest when on the opposite side of the sun from Earth, at which time almost all its sunlit area can be seen.

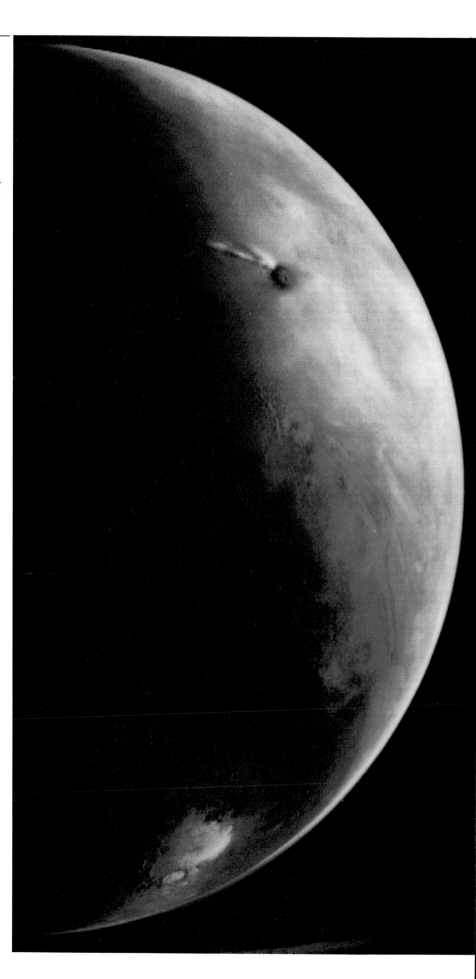

This photograph of Mars was taken by Viking 2 in 1976. Two wisps of clouds, thought to consist mostly of water vapor, can be seen. Situated near the top and at the bottom of the photograph, the clouds are next to dark spots, which are probably giant volcanoes. Between these two features and slightly to the right is a giant canyon scratched into the planet's surface.

60

MARS

No planet has captured human imagination more than Mars. The ancient Romans, awed by its reddish glow in the night sky, named the planet for their god of war. In more recent times, the red planet has been the favorite target of speculators and science-fiction writers who imagined vast civilizations there—the cities of Martians.

Mars occasionally comes to 48.7 million miles (78.4 million kilometers) from the Earth, closer to us than any planet except Venus. During these close approaches, anyone with a modest-sized telescope can see faint

Giovanni Schiaparelli (1835–1910) was an Italian astronomer who first detected dark lines on the surface of Mars.

This composite map, based on sketches made by Schiaparelli in June, 1888, shows the canali (channels) he observed on the surface of Mars.

patches of dark gray on the pinkish surface. Often, an observer can glimpse patches of white at the planet's poles; these are icecaps that resemble Earth's. But our view of Mars has always been a frustrating one, for it never shows us more than slight detail.

CANALS ON MARS?

In 1877, when Mars made an especially good showing, Italian astronomer Giovanni Schiaparelli turned his telescope on the planet and reported seeing dark lines, which he called *canali*, on the surface. Schiaparelli said nothing of their origin; canali simply means *channels* in Italian. But somehow the word was translated into English as *canals*. If there were canals on Mars, then someone must have built them—or so thought Boston-born astronomer Percival Lowell.

Lowell was convinced that the canals were the work of a Martian civilization, used to transport water from the polar icecaps to their cities in the Martian deserts. Working at his observa-

tory in Flagstaff, Arizona, he counted over 500 canals. Lowell died in 1916 without ever knowing if his Martians were real.

When bigger telescopes like the 200-inch (500-centimeter) reflector at Palomar Observatory in California were built, astronomers could see more detail on Mars than Lowell had seen. They found no evidence of canals or any other signs of civilization. But that didn't put an end to dreams of Martian life. A number of scientists interpreted the dark patches on Mars as vegetation, claiming to see changes in plant cover with the Martian seasons. Not until 1965 did the dreams of Mars as an Earthlike world finally die. That year a robot explorer called *Mariner 4* sped past Mars and radioed to Earth the first close-up pictures of its surface.

THE MARINERS TO MARS

Had Lowell been around to see *Mariner 4*'s pictures, he would have been surprised and perhaps heartbroken. Instead of civilizations and canals, only a barren crater-pocked surface was seen. Mars looked no more inspiring than the moon. But *Mariner 4* had only glimpsed a tiny portion of the planet. Perhaps there might be more interesting sights on other parts of Mars. In 1971, scientists sent another automated explorer, *Mariner 9*, into orbit around the red planet for a better look.

Unfortunately, *Mariner 9* arrived during the height of a planet-wide dust storm. Such storms are common on Mars, but this one was especially severe. For weeks, the entire planet was hidden by a shroud of wind-blown dust. Finally, in late 1971, the dust began to clear.

The first features to appear amid the haze were four dark spots. Later pictures showed that the spots were in fact craters.

More dust settled, and it became clear that these craters were perched atop enormous mountains. Geologists realized the mountains were four huge volcanoes. The largest, named *Olympus Mons* (after the dwelling place of the gods of Greek legends), towers 15 miles (25 kilometers) above the surface, nearly three times as high as Mount Everest. Lava once poured from its summit crater, which is some 50 miles (80 kilometers) across. But today, there are no signs of activity from any of the Martian volcanoes. They may be dormant, or they may have died out millions or even billions of years ago.

More surprises followed. East of the four volcanoes, a vast canyon almost as long as the continental United States cut through the Martian crust. Geologists named the canyon *Valles Marineris*, after the spacecraft that showed us Mars's true nature.

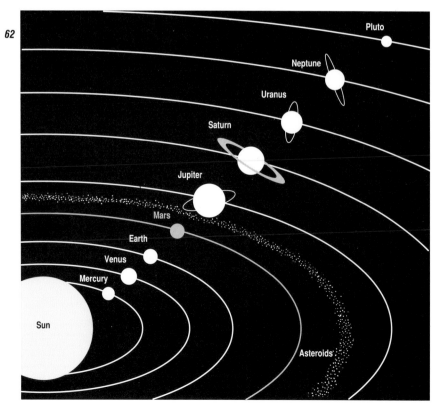

62

Mars at a glance

Mars, shown in orange in the diagram, is the next planet beyond the earth.

Distance from sun:
Shortest—128.4 million miles (206.6 million kilometers); *Greatest*—154.8 million miles (249.2 million kilometers); *Mean*—141.6 million miles (227.9 million kilometers).

Distance from earth:
Shortest—48.7 million miles (78.4 million kilometers); *Greatest*—248 million miles (399 million kilometers).

Diameter: 4,223 miles (6,796 kilometers).

Length of year: About 1 earth-year and 10.5 months.

Rotation period: 24 hours and 37 minutes.

Temperature: −225° to 63° F. (−143° to 17° C).

Atmosphere: Carbon dioxide, nitrogen, argon, oxygen, carbon monoxide, neon, krypton, xenon, and water vapor.

Number of satellites: 2.

(Planets and orbital distances are not to scale.)

Like the moon, Mars has thousands of impact craters of all sizes and a number of large impact basins filled with lava plains. And the crust of Mars—like the moon's, Mercury's, and Venus'—does not appear to be broken up into plates like Earth's. Strangely, Mars seems to be split into two halves—a northern region of volcanic plains and a southern region of crater-pocked uplands. Scientists have yet to explain this interesting phenomenon.

What of the light and dark markings seen from Earth? To no one's surprise, *Mariner 9* found that Mars is covered not with vegetation but with deposits of fine dust. These show up as bright areas in our telescopes. In some places, this dust has been swept away by the swift Martian winds to reveal the darker rocks beneath, accounting for most of the dark markings.

The handiwork of Martian winds—ranging from vast dune fields and streaklike deposits in the shelter of craters to occasional dust storms and dust devils—is seen over much of the planet. Over billions of years,

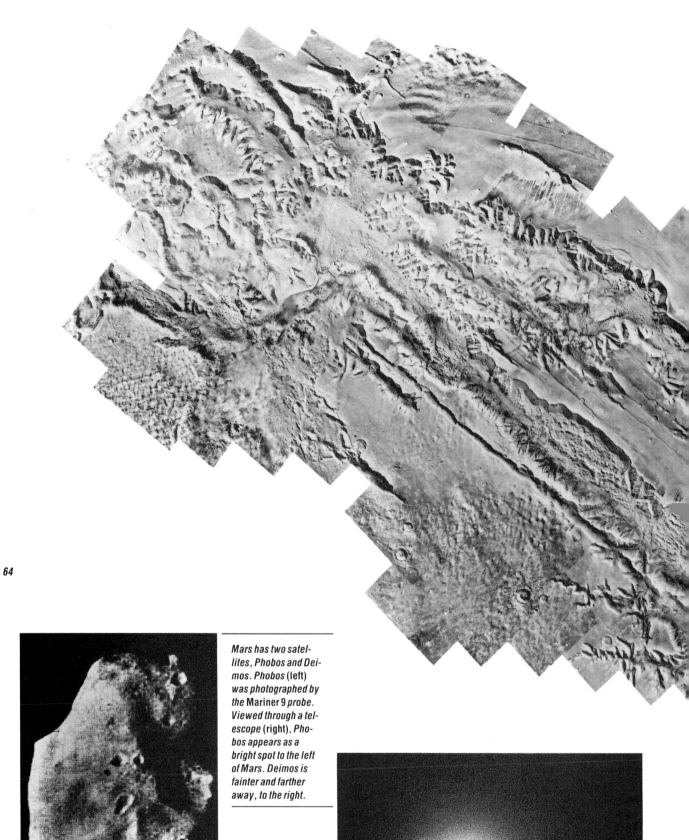

Mars has two satellites, Phobos and Deimos. Phobos (left) was photographed by the Mariner 9 probe. Viewed through a telescope (right), Phobos appears as a bright spot to the left of Mars. Deimos is fainter and farther away, to the right.

the erosive action of wind-blown dust has softened crater rims, mountains, and other landforms. But despite such erosive action, much of the ancient terrain has survived.

THE ANCIENT RIVERS OF MARS

One of *Mariner 9*'s most interesting discoveries was a set of channels (unrelated to Schiaparelli's channels) that look exactly like dry riverbeds seen on Earth. Scientists concluded the ones on Mars must surely have been carved by running water. Yet, no signs of liquid water are to be found on Mars today. In fact, the Martian atmosphere is so thin—with a surface pressure averaging about 100 times less than that at sea level on Earth—that a drop of water placed on Mars' surface would evaporate in seconds.

Furthermore, temperatures on Mars are usually far below the freezing point of water. At noon on a summer day at the Martian equator, the uppermost layer of dust might warm to 63° F. (17° C). But most of the time, Mars is far colder. At the poles, temperatures drop to as low as −225° F. (−143° C), which is cold enough to freeze carbon dioxide, forming deposits of dry ice on the surface. The vast icecaps covering the poles are mixtures of frozen carbon dioxide and water ice.

Clearly, Mars was a very different place when the channels were formed. There must have been a far denser atmosphere, perhaps one made largely of carbon dioxide, which now comprises most of the thin Martian air. Perhaps over time, most of the carbon dioxide was incorporated into the crust, much like Earth's carbonate rocks. Scientists believe that a thicker atmosphere on Mars would have allowed water to flow in the channels.

Where is all of Mars' water now? Some of it is contained in the polar icecaps, but they don't account for enough to have carved the big channels. Many geologists think the rest could be frozen in the upper layers of Mars' crust. But no one will be certain until future spacecraft are sent to Mars with sensors to detect the buried ice, or until human explorers drill into the crust of Mars.

THE VIKINGS ON MARS

On July 20, 1976, a three-legged spacecraft called *Viking 1* descended through the thin atmosphere of Mars. As it neared the surface, the craft cast off its large, white parachute and fired three small retrorockets to slow its final few yards of descent. With a sudden bump, *Viking 1* came to a stop in the dust of Mars' *Chryse Planitia*—the Plain of Gold.

For scientists back on Earth, *Viking 1*'s landing brought something more valuable than gold—the first pictures and analyses made from the surface of Mars.

The Valles Marineris, *or Mariner Valleys, is a huge complex of canyons stretching for hundreds of miles across the surface of Mars. More than 100 TV images taken by* Viking 1 *were used to form this composite photomosaic.*

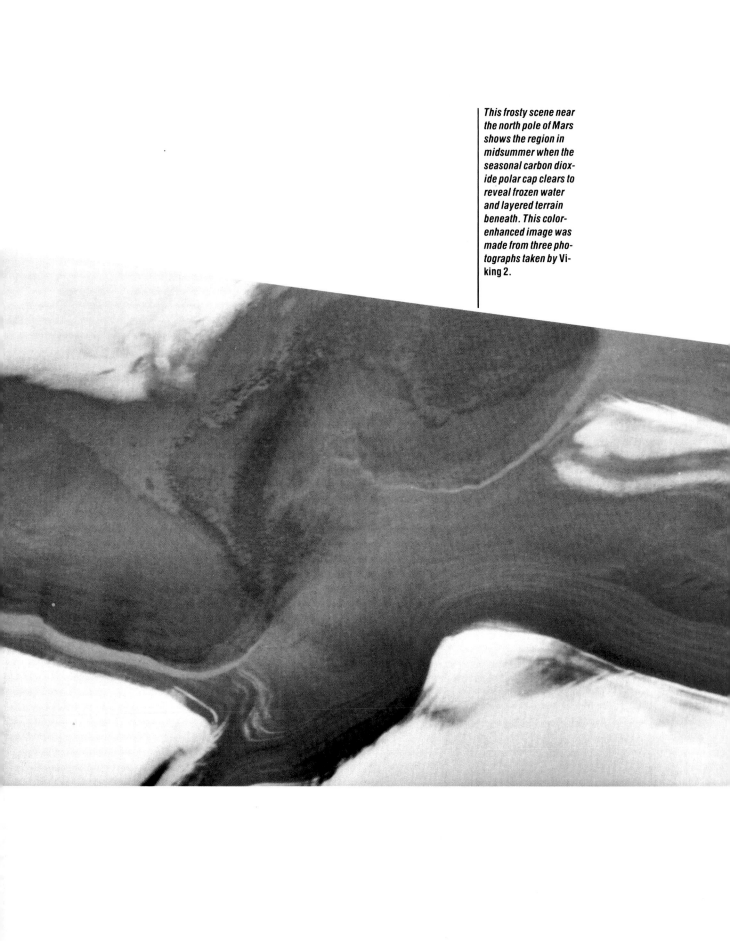

This frosty scene near the north pole of Mars shows the region in midsummer when the seasonal carbon dioxide polar cap clears to reveal frozen water and layered terrain beneath. This color-enhanced image was made from three photographs taken by Viking 2.

The images showed thousands of rocks strewn from horizon to horizon on a gently rolling plain. Dunelike deposits of fine dust lay only a few hundred feet from the lander. On September 3, 1976, *Viking 2* set down at Mars' *Utopia Planitia* (Utopian Plains) amid hundreds of rocks. On the distant horizon, the rim of an impact crater broke the monotony of an otherwise featureless plain.

The *Vikings* operated for more than two years on Mars, sending back a harvest of data on their surroundings, including weather reports and chemical analyses of the dust. Color-enhanced pictures showed that the yellowish-brown soil appears reddish-brown in the Martian atmosphere. The chemical analyses revealed that the color is caused by oxidized iron—otherwise known as rust. Because fine dust is suspended in the atmosphere, the Martian sky looks pink.

Along with cameras, weather instruments, and soil analyzers, the *Vikings* carried a very special instrument—a biology laboratory only 1 cubic foot (0.03 cubic meter) in size. With it, Viking scientists hoped to answer the nagging question of whether there is life on Mars. With no water available and no protection from the sun's deadly ultraviolet rays, Martian life would have to be unusually hardy. But the possibility could not be ruled out.

Each *Viking* used its mechanical arm to scoop up some of the soil and dump it into the biology instrument, where three separate tests were performed. One involved adding water to the dust and waiting to see what happened. Even a weak nutrient broth was injected into one sample, in hopes that any microbes living in the soil would multiply and their metabolic wastes be detected. Each of the tests gave peculiar results that almost mimicked the characteristic activity of microbes on Earth. But

when all the data were in, scientists concluded that highly reactive minerals, not Martian microbes, were responsible. If there is life on Mars, it eluded the Viking probes, but the question is still open in the minds of some hopeful scientists.

MISSION: MARS

In the early 1990's, NASA plans to launch the *Mars Observer*, which will orbit Mars and examine its surface, atmosphere, and climate. And in the late 1990's, NASA hopes to launch the *Mars Rover*, which would spend 5 months on Mars gathering samples of soil, rock fragments, and subsurface material. The mission will demonstrate technologies required for manned missions, and it may also involve international cooperation.

The Soviet Union also has plans for unmanned Mars exploration. In July, 1988, *Phobos 1* and *2* were launched toward Mars and the Martian moon Phobos. Communication broke down with both spacecraft at an early stage in the mission but not before a great deal of scientific data was gathered. Undeterred by this partial failure, Soviet scientists plan to send a land-rover to Mars in 1994 that will be equipped with television cameras and drills for analyzing the planet's surface. Cosmonauts are also currently testing their ability to endure long periods in space. Their goal could be preparation for a manned mission to Mars.

THE NEW MARTIANS

The Mars of fact turned out to be very different from the Mars of imagination. It is a less romantic world, perhaps, but no less interesting. Mars will continue to draw our attention. Someday in the not too distant future, astronauts will probably visit Mars to help solve more of

the red planet's mysteries. They will search for water, and they will search for life. Eventually, they will set up a permanent outpost on Mars, a way station for voyages farther out into the solar system. Then, there *will* be Martians, after all.

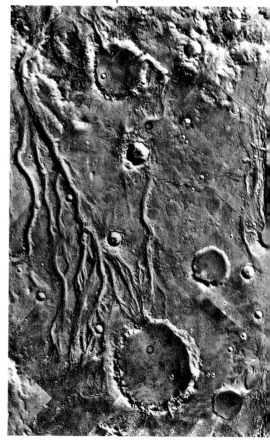

This photograph, taken by a Viking spacecraft, shows water "channels" on Mars. The channels seem to support the view of some scientists that large quantities of water once flowed on Mars's surface.

The Size and Color of Asteroids

Vesta

Hektor

Thetis

Nysa

Eunomia

Ceres

Davida

Laetitia

Psyche

Interamnia

Phobos

ASTEROIDS

Shown to scale against a portion of the moon, asteroids vary in size and shape, from tiny, potato-shaped Nysa to huge, spherical Ceres. Phobos, lower left, a captured moon of Mars, may also be an asteroid.

In the late 1700's, astronomers began searching for a "missing planet." According to Bode's law, a formula that accurately predicted the distances of the known planets from the sun, there was a gap between Mars and Jupiter where a planet ought to be. However, no planet was known to be there. But on the first night of 1801, an Italian astronomer named Giuseppe Piazzi discovered a small world, only about 600 miles (960 kilometers)

across, which was subsequently named Ceres. It was exactly where Bode's law said it should be: 257 million miles (414 million kilometers) from the sun. Apparently, the missing planet had been found. But in following years, more small bodies were found in the same part of the solar system. Astronomers called these bodies asteroids.

Where did the asteroids come from? Could it be that they were fragments of a much larger world, torn apart by an explosion or some other disaster? For many years, this was the accepted explanation for the asteroids' origin. But today, astronomers believe that asteroids are leftover debris from the formation of the solar system.

Today, the orbits of more than 3,000 asteroids are known. They range in size from Ceres, the largest asteroid, to tiny objects perhaps no bigger than grains of dust. Most asteroids orbit the sun between Mars and Jupiter in a doughnut-shaped zone called the Asteroid Belt. Dozens more have been discovered in the inner solar system on paths that cross Earth's orbit. These are called Apollo asteroids. And two swarms of asteroids share an orbit with the giant planet Jupiter. This last group is the Trojan asteroids.

WHAT ARE ASTEROIDS LIKE?

Asteroids probably resemble Phobos and Deimos, the two moons of Mars. Scientists believe Phobos and Deimos may be asteroids that were captured by Mars's gravity. Phobos and Deimos are irregular, battered chunks of rock. Their surfaces

The Willamette meteorite was found in Oregon in 1902. It is the largest meteorite ever found in the United States, measuring almost 118 inches (300 centimeters) in length and weighing about 15.5 short tons (14 metric tons).

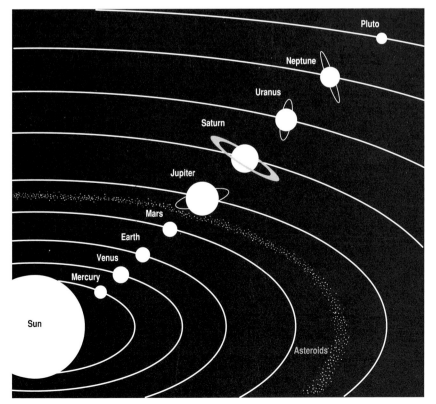

Asteroids at a glance

Asteroids, shown in orange in the diagram, occupy the space between Mars and Jupiter.

Distance from sun: Most asteroids are between the orbits of Mars (141.7 million miles, 228 million kilometers) and Jupiter (483.7 million miles, 778.4 million kilometers). Ceres, the largest asteroid, is at 257 million miles (414 million kilometers).

Number: at least 30,000. The orbits of over 3,000 are known.

Size: ranging from the largest, Ceres, about 600 miles (960 kilometers) in diameter, to less than 1 mile in diameter. About 30 asteroids have diameters greater than 120 miles (192 kilometers).

Shape: ranging from nearly ball-shaped to highly irregular.

Composition: mostly rock or a mixture of rock and metal.

(Planets and orbital distances are not to scale.)

are heavily cratered from billions of years of collisions with other objects.

Scientists think many asteroids have been broken into pieces by violent collisions with one another. In some cases, the fragments fell back together again, attracted by their own gravity, to form a cloud of rubble orbiting the sun. Other asteroids might have smaller asteroids orbiting around them. Still others might come in pairs that perhaps touch as they slowly rotate around each other.

Most asteroids are too small to be seen even with powerful optical telescopes. Yet, scientists have learned something about what asteroids are like by analyzing the light reflected off their surfaces. By breaking down the light with a device called a spectrograph, astronomers have found that asteroids have different surface compositions. Some are made of rock, and others are mixtures of rock and metal. A rare few even have lava plains on their surfaces, like the lunar maria.

METEORITES

Scientists may already have samples of asteroidal material, even though no spacecraft has ever visited an asteroid. These samples may be fragments blasted off the surface of asteroids during violent collisions. In such collisions, fragments of rock or metal—called meteoroids—shoot into space in all directions.

Some enter Earth's atmosphere, where friction with the air makes them glowing hot. We see them as bright streaks of light, called meteors or falling stars. Meteoroids that reach Earth's surface before burning up are called meteorites.

What do meteorites tell us about asteroids? Some meteorites, called chondrites, may be chunks of the original material that formed the sun and planets. Certain chondrites—called carbonaceous chondrites—contain organic molecules, the chemical building blocks of life, and are 4.6 billion years old—as old as the solar system itself. Phobos and Deimos, the two moons of Mars, are probably made of carbonaceous chondrite material.

ACTIVE ASTEROIDS

Not all asteroids have remained unchanged since their formation. Some became so hot inside that their interiors melted, forming metal cores and dense, rocky mantles like Earth's. Some of these asteroids must have been shattered to bits by collisions with other asteroids, because we have found meteorites made of iron or mixtures of iron and rock—samples of the cores and mantles of these fragmented asteroids.

Some scientists have proposed mining asteroids for their metals—even a small asteroid could supply many tons of metal for industries on Earth and in space.

Some meteorites are made of basalt, a volcanic rock also found on Earth and the moon. The asteroids that spawned these basaltic meteorites must have been volcanically active at some time in their history. On

these asteroids, molten rock erupted to the surface and formed lava plains. Some of the basaltic meteorites have tentatively been traced to Vesta, one of the largest asteroids.

ASTEROID EXPLORATION

Both the United States and the Soviet Union plan to send spacecraft to explore asteroids. The United States, in fact, plans to launch the *Comet Rendezvous Asteroid Flyby (CRAF)* in August, 1995. *CRAF* will be equipped to visually map an asteroid called Hedwig before reaching and flying alongside Comet Wild 2. Close-up pictures of an asteroid would tell much about the processes that have affected it since its formation. But most valuable of all would be a piece of an asteroid brought back to Earth for study. With pictures and other data on the asteroid it came from, scientists might learn more from the sample than from meteorites, whose origins are not known for certain. If asteroids really are the leftovers from

Some 50,000 years ago, a meteorite slammed into Earth and formed the Meteor Crater in northeastern Arizona. It is about 4,150 feet (1,265 meters) across and 570 feet (174 meters) deep.

On August 10, 1972, a meteor skipped through Earth's atmosphere and continued out into space. This picture of the meteor was taken in Grand Teton National Park in Wyoming.

the birth of the solar system, they may have much to tell us about how the solar system came to be.

CLOSE ENCOUNTER

A huge asteroid on a celestial collision course with Earth whizzed past our planet on March 23, 1989, coming closer than any other such heavenly body in 52 years. If the giant rock had hit Earth, the impact v uld have equaled the explosion of thousands of H-bombs and would probably have killed millions of people. Of course, this "close call" was only close in a relative sense—it was 450,000 miles (720,000 kilometers) away, about twice the distance between the Earth and the moon. Still, most agree that the near-encounter was too close for comfort. Many scientists believe that it was an equivalent-sized asteroid that caused the extinction of dinosaurs 65 million years ago.

Two of the four Galilean satellites are shown orbiting above Jupiter's multicolored clouds. The Great Red Spot can be seen on the lower left portion of the planet.

JUPITER

Jupiter—named for the king of the gods of ancient Rome—is the mightiest of planets in the solar system. It is the largest planet, some 88,846 miles (142,984 kilometers) across, more than 10 times the size of Earth. Jupiter is so big that if it were hollow, all the other planets would fit inside it.

Jupiter is more like the sun than like Earth. Like the sun, Jupiter is mostly hydrogen, with a small amount of helium. In addition, the clouds of Jupiter contain water, methane, ammonia, and other compounds of nitrogen, hydrogen, and sulfur, all of which are thought to have existed in the earliest atmosphere of Earth. Scientists think that if Jupiter had been about 70 times more massive, it would have been a star. As it is, Jupiter is something halfway between a star and a solid planet.

THE CLOUDS OF JUPITER

Through even a small telescope, Jupiter displays distinct light and dark bands that cross the planet parallel to its equator. Jupiter has no solid surface—these bands are clouds. The planet's rapid rotation (once every 10 hours at the equator) and very high winds cause these clouds to be drawn out into alternating light and dark bands, which astronomers call belts and zones. The belts and zones race around the planet at high speeds, like the jet streams in Earth's atmosphere.

It is difficult to see any detail in Jupiter's clouds, since at its closest the planet is some 391 million miles (626 million kilometers) from us. But spacecraft have flown past Jupiter and taken close-up pictures, which show that its turbulent clouds are constantly changing. The composition of the clouds is not definitely known, but they appear to be made up of three layers. The bottom layer may be composed of water in the form of ice crystals and possibly droplets of liquid water. The middle layer may contain a compound

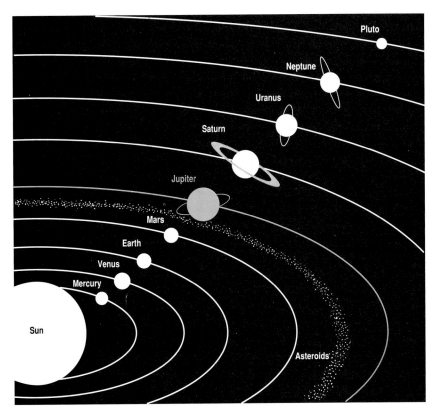

Jupiter at a glance

Jupiter, shown in orange in the diagram, is the fifth closest planet to the sun.

73

Distance from sun: *Shortest*—460.2 million miles (740.6 million kilometers); *Greatest*—507.1 million miles (816 million kilometers); *Mean*—483.6 million miles (778.3 million kilometers).

Distance from earth: *Shortest*—390.7 million miles (628.8 million kilometers); *Greatest*—600 million miles (970 million kilometers).

Diameter: 88,846 miles (142,984 kilometers).

Length of year: About 12 earth-years.

Rotation period: 9 hours and 55 minutes.

Average temperature: −234° F. (−148° C).

Atmosphere: Hydrogen, helium, methane, ammonia, carbon monoxide, ethane, acetylene, phosphine, water vapor.

Number of satellites: 16.

of ammonia and hydrogen sulfide, and the uppermost layer seems to be ammonia ice crystals.

Perhaps the most fascinating of Jupiter's features is the Great Red Spot, a gigantic storm three times the size of Earth that has been raging for at least a few hundred years. It has changed in appearance over the years, becoming very prominent at some times and nearly disappearing at others.

No one knows what has kept the Great Red Spot going for so long, but it may get its energy from the atmospheric jets around it, like a ball bearing spinning between two conveyor belts. Motion pictures made from *Voyager* spacecraft images show that the Great Red Spot spins counterclockwise, the same direction as hurricanes in Earth's Northern Hemisphere. Perhaps by studying the Great Red Spot, meteorologists will learn more about storms in our own atmosphere.

JUPITER'S RADIATION

Jupiter does something very strange: it radiates about twice as much heat as it receives from the sun. Scientists think the heat is still escaping from the time Jupiter was formed.

Most of the contraction occurred long ago. Early in its history, Jupiter probably glowed red hot. It would have been an awesome sight in the skies of its newly formed satellites. Today, Jupiter is probably still shrinking at a slow rate that is too small to measure.

Jupiter's clouds are very cold, about −234° F. (−148° C). But deep within the planet's interior, temperatures may rise to thousands of degrees above zero. Somewhere between the extremes, perhaps below the visible clouds, the temperatures

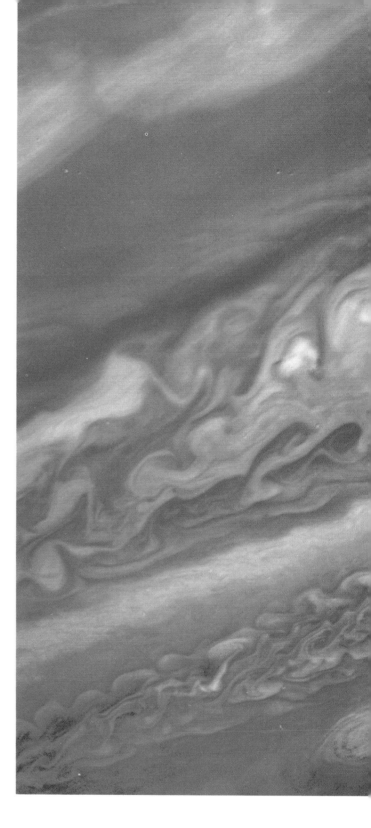

This photograph of Jupiter's Great Red Spot was taken by Voyager 2 in July, 1979. This photo is different from the one taken of the same region by Voyager 1, showing that the dense layer of clouds around Jupiter is constantly changing.

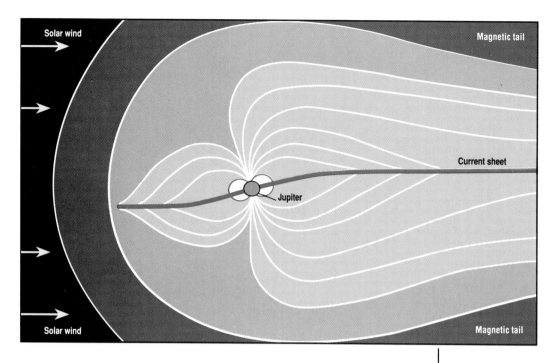

Solar wind

Magnetic tail

Current sheet

Jupiter

Solar wind

Magnetic tail

Jupiter's invisible magnetosphere is many times larger than the planet itself. The magnetosphere is molded by the solar wind into a magnetic tail that stretches away from the sun. The current sheet is a thin band of electrically charged particles.

may be similar to those found over the surface of Earth and would be comfortable for life as we know it. It has been suggested that perhaps a form of life exists there, floating within the lower atmosphere.

Some scientists believe Jupiter's extra heat is the energy source for the planet's swirling clouds. Others say Jupiter's clouds, like Earth's, get their energy from sunlight. To resolve the question, scientists are eagerly awaiting the findings that will be revealed when the unmanned *Galileo* spacecraft lands an instrumented probe on Jupiter in 1995.

Besides heat, Jupiter's magnetosphere traps high-energy particles, such as hydrogen, helium, oxygen, and sulfur ions. These particles form belts of intense radiation in Jupiter's magnetic field.

INSIDE AND OUTSIDE JUPITER

No one knows what the interior of Jupiter is like. Beneath the clouds, perhaps 600 miles (960 kilometers) down, there may be a layer of hydrogen in liquid form, due to the intense pressures and low temperatures. Deeper in the interior, pressures may squeeze the liquid hydrogen so much that it behaves more like metal than liquid. Finally, there might be a solid core roughly the size of Earth at Jupiter's center, composed of iron and other metals and rock.

When the *Voyager 1* spacecraft flew past Jupiter in 1979, it discovered that the planet is encircled by a thin ring like Saturn's, only much smaller, roughly 4,000 miles (6,400 kilometers) wide. The ring is made of microscopic, dark particles of unknown origin.

This color picture of Io was taken by Voyager 1. The red and orange colorations are probably surface deposits of sulfur compounds, salts, and possibly other volcanic sublimates. The dark spot with the irregular radiating pattern near the bottom of the picture may be a volcanic crater with radiating lava flows.

This color reconstruction of Ganymede's surface was made from a series of pictures taken by Voyager 2. It shows part of dark, densely cratered block which is bound on the south by lighter, less cratered grooved terrain. The circular features are craters.

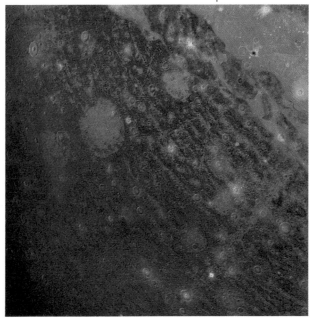

This color image of Europa was taken by Voyager 2. The complex patterns on its surface suggest that a previous icy surface was fractured, and that the cracks filled with dark material from below. Very few impact craters are visible on the surface, suggesting that active processes on the surface are still modifying Europa.

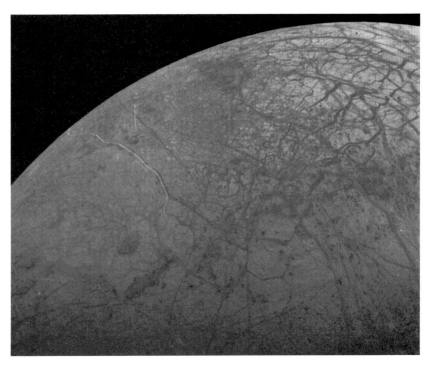

When Galileo turned his telescope on Jupiter in 1610, he discovered that the planet has four large satellites. Today, astronomers know that at least 16 moons orbit Jupiter, forming a kind of miniature solar system. These satellites probably condensed out of a cloud of gas and dust orbiting Jupiter, just as the planets condensed out of a similar cloud surrounding the sun. Jupiter's cloud contained some dust and a great deal of water; today, its satellites are mixtures of rock and ice. Most of the satellites are small, insignificant chunks about which little is known. But Galileo's four discoveries—called the Galilean satellites—are fascinating worlds, small planets in their own right.

Callisto

The outermost of the Galilean satellites, Callisto, is a ball of ice and rock some 3,000 miles (4,800 kilometers) across, making it about the same size as the planet Mercury. Callisto's entire surface is covered with impact craters. Billions of years of bombardment by cosmic debris are visible on its dark, icy crust. Some of Callisto's craters are surrounded by bright rays of ice blasted out of subsurface layers. There is a huge, ringed impact basin called Valhalla, possibly the largest impact crater in the solar system. Researchers believe Callisto's icy crust flowed very slowly, smoothing out major topographic features over billions of years. Valhalla and other large craters on Callisto have flattened out completely, leaving only light colored circular patches.

Ganymede

Moving inward from Callisto, we encounter Ganymede, the largest moon in the solar system. Its diameter is approximately 3,300 miles (5,300 kilometers). Like Callisto, Ganymede is probably half frozen water and half rock.

Seen from a great distance, Ganymede displays dark patches set against a light background. Close-up photographs show that the dark areas resemble Callisto's heavily cratered surface. But the dark areas on Ganymede are surrounded by lighter colored terrain, crisscrossed by peculiar ridges and grooves. The grooves seem to be places where Ganymede's crust was fractured, allowing water to pour out from the interior.

Ganymede's grooved terrain—where blocks of ice have slid past each other—reminds scientists of Earth's crustal plates. Although the grooves are younger than Ganymede's dark areas, they are still very old, having formed perhaps 4 billion years ago. At that time, Jupiter is thought to have been much brighter and hotter than today. It probably glowed red hot. Ganymede, being closer than Callisto, felt the effects of Jupiter's heat more. As Jupiter cooled down, Ganymede's geologic activity slowly ceased.

Europa

The next moon in from Ganymede is Europa, 1,945 miles (3,112 kilometers) across. Europa contains a lower percentage of water than Ganymede and Callisto. This might be because Eu-

78

Jupiter's four Galilean moons are also the four largest of the planet's 16 known satellites: Io *(upper left);* Europa *(upper right);* Ganymede *(lower left); and* Callisto *(lower right).*

This photograph shows a special color reconstruction of one of the erupting volcanoes on Io discovered by Voyager 1.

ropa formed closer to Jupiter, where temperatures were higher and less water condensed.

Europa's icy surface is one of the brightest in the solar system. Strangely, the satellite is as smooth as a billiard ball—no large impact craters, mountains, or major relief of any kind has been found on its surface. It is likely that water has flowed onto the surface and frozen, covering everything with a smooth sheet of bright, fresh ice.

Europa is covered by a network of dark and light lines. Geologists think that at some time in Europa's history the crust cracked, in the same manner that sea ice on Earth does. Later, presumably, the cracks were filled in by water or other material from below, which froze and created the dark and light lines that we see today.

No one can say what lies beneath the icy crust of Europa, but there may be a deep ocean of water or slushy ice. What would keep this ocean from freezing? Scientists think massive Jupiter and the other Galilean satellites may constantly tug at Europa, causing it to flex. Such gravitational tuggings may have heated Europa's interior, melting some of its ice and creating a subterranean ocean. Some scientists have proposed that some form of life might be found in such an ocean if it exists. Most of Europa, however, is probably made of rocky material.

Io

Io, the innermost of the four Galilean satellites, is a bizzarre world about the size of Earth's moon. But it is nothing like the moon or any other planet. Instead of ice, Io is mostly rock. Heat from the infant Jupiter might have been enough to keep any water from condensing when this nearby moon formed.

Io is covered with active volcanoes. These volcanoes, according to many scientists, pour forth primarily molten sulfur and silicates. Io's volcanoes also spray umbrella-shaped plumes of sulfur dioxide ice crystals up to 437.5 miles (700 kilometers) or more into space, giving the satellite a thin sulfur dioxide atmosphere.

Io's surface is very cold, about −231° F. (−146° C), but at the mouths of its volcanoes, temperatures as high as 63° F. (17° C) have been recorded. Sulfur that pours out of the volcanoes onto Io's surface quickly solidifies in the cold of space.

When the *Voyager 1* spacecraft flew by Io in 1979, eight volcanoes were erupting at once. By comparison, no more than one or two volcanoes are likely to be erupting at any one time on Earth. Io is easily the most active world geologically in the solar system.

What powers Io's volcanoes? Io experiences strong tidal stresses caused by the gravitational forces of Jupiter and other moons, especially Europa. These tidal forces squeeze and unsqueeze Io in each orbit. Such flexing pumps energy into the interior of Io in the form of heat. As a result, widespread and recurrent volcanic eruptions occur.

Io's volcanoes pour out so much material that the surface of this moon is constantly changing. In fact, one could say that Io is continually turning itself inside out. No impact craters of any kind have been spotted on Io. They must lie buried deep beneath Io's present surface.

You would not want to go to Io without all kinds of protective gear. Apart from the extreme cold, Io orbits inside the deadly radiation belts associated with Jupiter's magnetic field. But if you could somehow safely visit this sulfurous moon, you would behold an awesome sight—Jupiter, as seen from Io, appears some 40 times larger than our moon looks from Earth. For the time being, however, scientists will probably send unmanned spacecraft to explore Io's fantastic surface.

Captured Asteroids?

Little is known of Jupiter's 12 other moons. Four small bodies circle Jupiter inside Io's orbit. The largest of these is Amalthea, a potato-shaped moon probably made of rock. Amalthea has a reddish surface that may be coated with sulfur. Two other groups of moons orbit beyond Callisto. Four are located about 7,000,000 miles (11,200,000 kilometers) from Jupiter, and another 4 are about 14,000,000 miles (22,400,000 kilometers) out. This last group circles Jupiter in the opposite direction compared to the other 12 satellites. And all 8 of these outer moons have orbits that are highly inclined to Jupiter's equator. They may be asteroids captured by Jupiter's gravity, or they may be fragments of larger moons that broke apart.

Scientists will learn more about Jupiter and its miniature solar system when the *Galileo* spacecraft, launched on October 18, 1989, begins orbiting Jupiter in 1995. *Galileo* will continue orbiting for two years as it studies the planet's atmosphere and makes repeated close flybys of its four Galilean satellites.

SATURN

This enhanced-color photograph of Saturn was taken by **Voyager 2** *in July, 1981, when the spacecraft was 21 million miles (34 million kilometers) from the planet. Two moons, Rhea and Dione, can be seen below the planet.*

Because of its majestic system of rings, Saturn has earned a place as one of the most beautiful planets in the solar system. Even a small telescope will show the rings surrounding the tan colored ball of Saturn itself. What are these rings? In a telescope, they look like solid sheets, but they are actually composed of billions of particles orbiting Saturn. The perimeter of the outer-

most ring may measure up to 272,400 miles (435,800 kilometers). It extends outward from Saturn as far as 46,000 miles (74,000 kilometers). All the rings are relatively thin, varying in thickness from about 660 feet (198 meters) to 9,800 feet (2,940 meters).

Most of the ring particles range in size from small chunks only an inch or so across to boulders the size of a house. The ring particles are apparently snowballs made of ice, but there may be other matter, such as dust, hidden inside them.

THE RINGS OF SATURN

There are seven major parts to the rings around Saturn. From Earth, astronomers can see three main parts of the rings that are called, from outermost to innermost, ring A, ring B, and ring C. Of these, A is fairly dark, B is the brightest, and C is semitransparent and more difficult to see. A fourth ring, called ring D, is too faint to be seen from Earth; it lies inside ring C. Rings F, G, and E lie outside ring A.

When the *Voyager 1* spacecraft sent back close-up pictures of Saturn in 1980, astronomers were amazed to find that each of the major rings is composed of thousands of narrow ringlets, or bands, giving the appearance of grooves on a phonograph record. *Voyager 1* also observed several narrow rings outside ring A. All in all, Saturn's ringlets probably number in the thousands.

A 3,100-mile (4,960-kilometer) gap, called the Cassini division, separates ring A from ring B. In addition, there is a gap within the A ring called the Encke division, also named the Keeler gap. Numerous small gaps exist throughout Saturn's rings.

BIZARRE DISCOVERIES

Most of the rings follow simple, nearly circular paths around Saturn. But a few are quite peculiar. Some rings are elliptical compared to the rest. Others have wavy edges. But the strangest

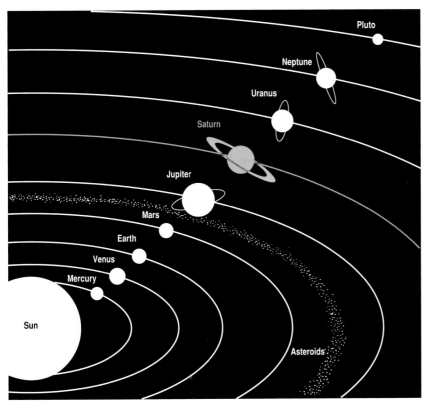

Saturn at a glance

Saturn, shown in orange in the diagram, is the sixth closest planet to the sun.

Distance from sun: *Shortest*—838.8 million miles (1,349.9 million kilometers); *Greatest*—937.6 million miles (1,508.9 million kilometers); *Mean*—888.2 million miles (1,429.4 million kilometers).

Distance from earth: *Shortest*—762.7 million miles (1,277.4 million kilometers); *Greatest*—1,030 million miles (1,658 million kilometers).

Diameter: 74,898 miles (120,536 kilometers).

Length of year: About 29.5 earth-years.

Rotation period: 10 hours 39 minutes.

Temperature: −288° F. (−178° C).

Atmosphere: Hydrogen, helium, methane, ammonia, ethane, and phosphine (?).

Number of satellites: 17.

(Planets and orbital distances are not to scale.)

An artist's conception of the view from within Saturn's rings shows chunks of ice in various sizes. The orange sphere to the left is Saturn, and the bright light to the right is the sun.

ring is F, photographed by the Voyager probes. This ring appears to be made of several separate strands, each less than 16 miles (25 kilometers) wide, that seem to be braided around each other. Scientists are mystified by ring F's strange braids, which cannot be explained by the normal laws of orbital motion.

Voyager 1 found two small moons circling Saturn on either side of ring F. Their gravitational attraction for particles keeps the ring confined to a narrow band. Perhaps the moons also create a gravitational disturbance in the ring's strands, causing their mysterious distortions. Scientists named the two moons the shepherding satellites. Another tiny moon orbits just outside ring A, and it appears to keep the ring's edge sharply defined.

Saturn's major satellites are far away from the edge of the rings, but they also influence the ring's particles with their gravity. Scientists are only beginning to understand the complex effects the moons have on the rings. The moons may be responsible for creating gaps in the rings, such as the Cassini and Encke divisions.

A final ring mystery, also photographed by *Voyager 1*, is dusky shadings that cross ring B like spokes on a wheel. The spokes travel around Saturn with the ring, changing in size and brightness as they continually emerge

Giovanni Domenico Cassini (1625–1712), the Italian-born French astronomer, was the first director of the Paris Observatory. His achievements include the discovery of four of Saturn's moons and the detection of a gap between two rings, which now bears his name: the Cassini division.

This color-enhanced photograph showing a section of Saturn's rings was taken by Voyager 2 in August, 1981. The rings can be seen to consist of thousands of thin ringlets.

This photograph was taken by Voyager 2 in August, 1981. It shows sporadically spaced "spokes" in the outer half of the broad B ring.

and fade away. As with so many features of Saturn's rings, these spokes are difficult to explain. One theory says they are made of particles that are held floating above the ring by electric forces and are swept along in Saturn's magnetic field as the planet rotates.

Astronomers are still trying to find out how Saturn's rings came to be. They know that the rings are located inside Saturn's Roche limit, which is the distance from the planet within which Saturn's tidal force would break up weak objects. The ring particles may have condensed from the same cloud of gas and dust that formed Saturn and its moons. Because they are inside Saturn's Roche limit, they never became joined into a single body. Or, they may be the remains of a body that broke apart after straying inside the Roche limit.

JUPITER'S FIRST COUSIN

What of Saturn itself? Like Jupiter, it is made of hydrogen and a small amount of helium. But Saturn is less dense than Jupiter. Its density is less than that of water, which means that if there were an ocean enormous enough to put Saturn in, the planet would float.

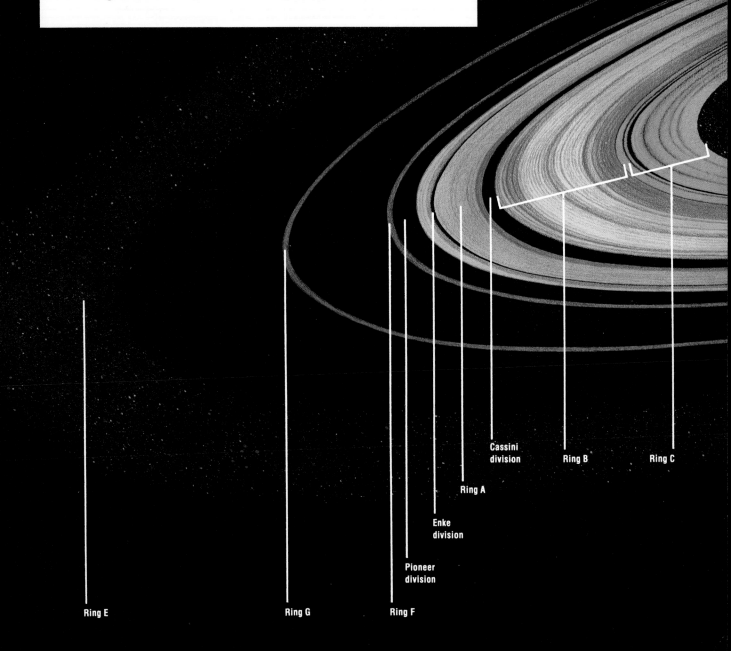

Cassini division

Ring B

Ring C

Ring A

Enke division

Pioneer division

Ring E

Ring G

Ring F

This cutaway view of Saturn shows what the interior is thought to be like. It also shows the major ring features surrounding the planet.

Outer core of ammonia, methane, and water

Liquid molecular hydrogen and helium

Liquid metallic hydrogen

Inner iron-rich core

Ring D

This color-enhanced image of Saturn's clouds shows light and dark bands that move rapidly around the planet. The swirling features within the bands may be atmospheric storms.

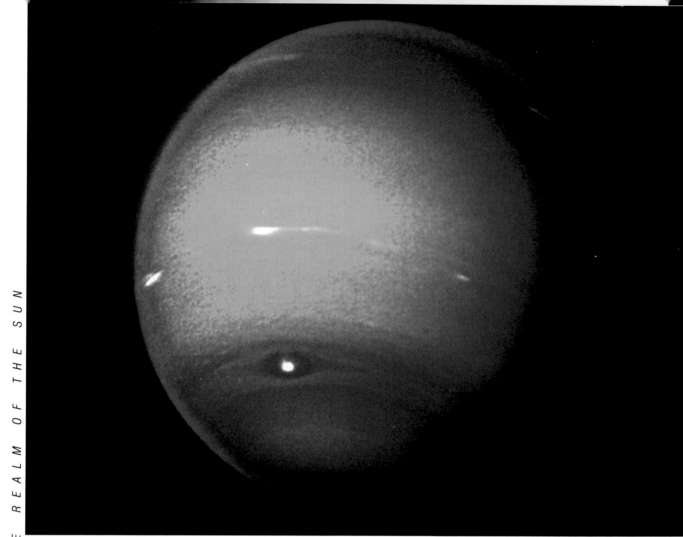

NEPTUNE

After Uranus was discovered in 1781, astronomers began to notice that its path around the sun did not conform to Kepler's laws of motion, which mathematically describe planetary orbits. The astronomers reasoned that perhaps some unknown planet lay beyond Uranus and was affecting Uranus' orbit with its gravitational pull.

In 1846, astronomers found Neptune, which lies roughly 2.8 billion miles (4.5 billion kilometers) from the sun—30 times farther than Earth. As it turns out, Neptune's gravity does not account for Uranus' motion. Astronomers still don't understand the cause.

Until recently, scientists believed that Neptune was Uranus' twin. But the images radioed to Earth on August 24, 1989, from *Voyager 2* upset many of these beliefs. Like Uranus, Neptune's atmosphere is composed of hydrogen and helium mixed with methane. Temperatures on both planets are similar, too. Neptune's surface averages a chilly −353° F. (−218° C). Unlike its neighbor, however, Neptune has turbulent winds sweeping its surface at the rate of 400 miles (640 kilometers) per hour. And a storm system churns counter-

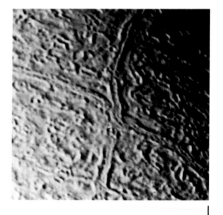

*This **Voyager 2** photograph reveals the mottled landscape of Triton. The long feature shown is probably a narrow down-dropped fault block.*

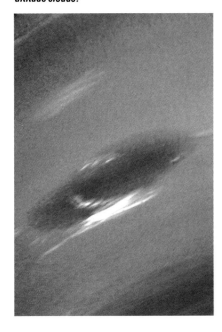

Voyager 2 *sent back to Earth this false-color image of Neptune (left). The red areas are a semitransparent haze covering the planet. The close-up below shows the storm system dubbed the Great Dark Spot, accompanied by high-altitude clouds.*

clockwise at 700 miles (1,120 kilometers) per hour. Scientists have dubbed this violent hurricane in Neptune's southern hemisphere the Great Dark Spot.

Radio emissions from Neptune had seemed to indicate a magnetic field twice the strength of Earth's. However, *Voyager 2's* sensing instruments revealed that Neptune's magnetic field tilts 50° from its axis of rotation, is only about as strong as Earth's field, and produces auroras spread over a wide area of the planet. Neptune is encircled by several complete rings as well as by a radiation belt similar to that around Earth. We also learned from *Voyager* that the planet has eight satellites, not two as previously believed.

Voyager 2 also flew by Neptune's largest moon Triton and discovered that it appears to be the coldest place in the solar system. Its surface temperature is about −400°F. (−206°C). Its diameter is approximately 1,700 miles (2,720 kilometers), about 400 miles (640 kilometers)

smaller than our moon. Triton's surface is a landscape of canyons, peaks, and craters, resembling the skin of a hugh cantaloupe, possibly caused by violent eruptions of ice volcanoes at its south polar cap. Its frigid temperatures, thin nitrogen-rich atmosphere, and low surface pressure support this new theory. Although Triton does not have a magnetic field, charged particles in Neptune's radiation belt appear to plunge into Triton's atmosphere and generate auroras at its equatorial plane.

Triton may once have been a planet in its own right, until it was captured by Neptune's gravitational pull. Its orbit around Neptune is circular, but tilted by 157°. Triton moves in a retrograde direction—opposite to Neptune's rotation, the only large moon in the solar system to do so. Nereid, another of Neptune's satellites, follows a long, narrow, elliptical orbit that is inclined by 29°. The other six moons all orbit near Neptune's equatorial plane.

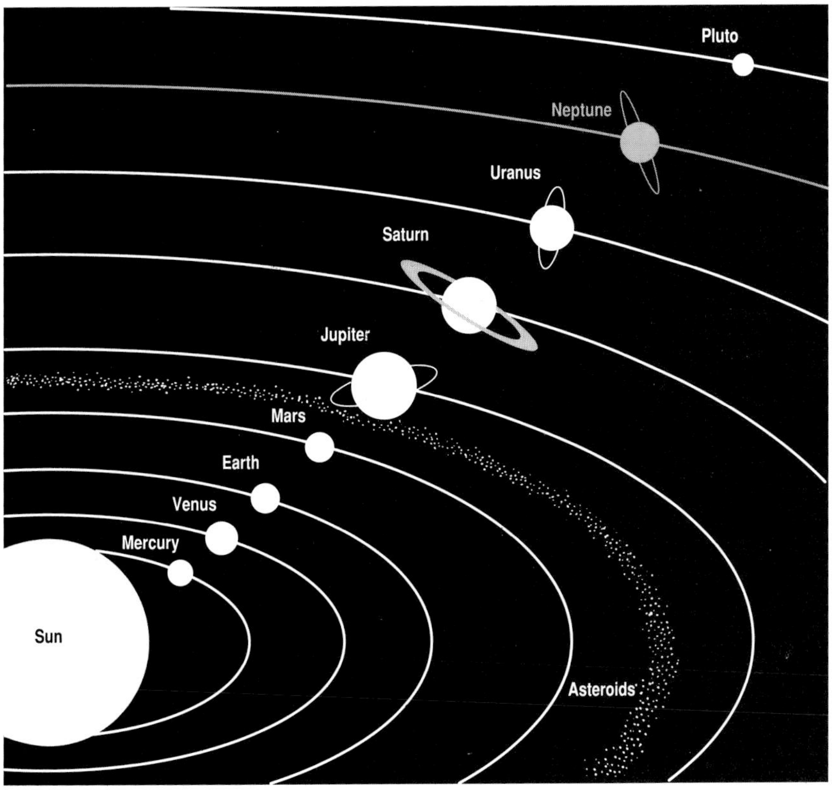

Neptune at a glance

Neptune, shown in orange in the diagram, is the eighth planet out from the sun. Every 248 years, Neptune passes outside the orbit of Pluto for a period of 20 years. During this time, Neptune is the farthest planet out from the sun.

Distance from sun: *Shortest*—2,774.8 million miles (4,465.6 million kilometers); *Greatest*—2,824.8 million miles (4,546.1 million kilometers); *Mean*—2,798.8 million miles (4,504.3 million kilometers).

Distance from earth: *Shortest*—2,700 million miles (4,350 million kilometers); *Greatest*—2,750 million miles (4,426 million kilometers).

Diameter: 30,775 miles (49,528 kilometers).

Length of year: About 165 earth-years.

Rotation period: About 16 hours and 11 minutes.

Temperature: −353° F. (−214° C).

Atmosphere: Hydrogen, helium, methane, and acetylene.

Number of satellites: Eight.

(Planets and orbital distances are not to scale.)

Pluto

Neptune

Uranus

Saturn

Jupiter

Mars

Earth

Venus

Mercury

Sun

Asteroids

PLUTO

Pluto is named for the ancient Roman god of the underworld. Averaging 3,666.2 million miles (5,900.1 million kilometers) from the sun, Pluto moves in a realm of near darkness and extreme cold—temperatures at Pluto's surface hover around $-369°$ F. ($-223°$C). From Pluto's surface, the sun would be a large, bright dot set against the backdrop of other stars.

Pluto was discovered in 1930 by Clyde Tombaugh, ending a search that was begun 25 years earlier by Percival Lowell. Lowell had begun looking for a *Planet X* when Neptune failed to

account for the motion of Uranus. Pluto follows an inclined path around the sun that is so elliptical that it brings the planet inside the orbit of Neptune. Pluto crossed the orbit of Neptune in 1979 and will not regain its status as outermost planet until 1999.

Pluto is only a faint smudge in the largest telescopes. Its size, mass, and composition are not known with certainty. Astronomers have determined that Pluto is less than a quarter the size of Earth. Several satellites of other planets are bigger than Pluto. Scientists have found evidence of methane frost and indications of a methane atmosphere on

An artist's conception of what Pluto's moon, Charon, might look like when viewed from Pluto's surface.

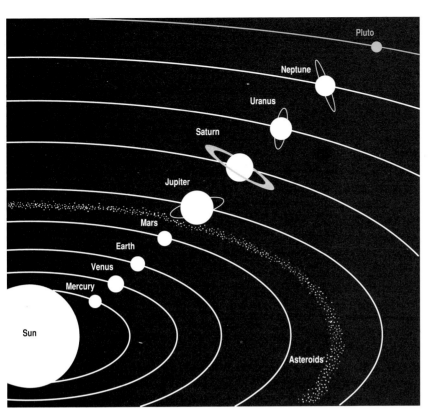

Pluto at a glance

Pluto, shown in orange in the diagram, is the farthest planet from the sun. Every 248 years, Pluto passes inside the orbit of Neptune for a period of 20 years. During this time, Pluto is the eighth planet out from the sun.

Distance from the sun: *Shortest*—2,749.6 million miles (4,425.1 million kilometers); *Greatest*—4,582.7 million miles (7,375.1 million kilometers); *Mean*—3,662.2 million miles (5,900.1 million kilometers).

Distance from the earth: *Shortest*—3,583 million miles (5,765.5 million kilometers); *Greatest*—4,670 million miles (7,516 million kilometers).

Diameter: 1,430 miles (2,300 kilometers).

Length of year: About 249 earth-years.

Rotation period: About 6 earth-days.

Temperature: About −387° to −369° F. (−233° to −223° C).

Atmosphere: Methane, nitrogen (?).

Satellites: 1.

(Planets and orbital distances are not to scale.)

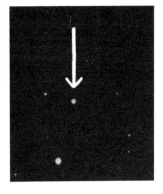
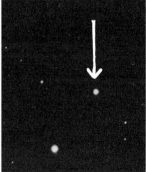

Even through powerful telescopes, Pluto looks like a dim star. Yet, these two photographs, taken a day apart, show Pluto's movement against the backdrop of distant stars.

Pluto, giving it a similarity to Neptune's moon Triton.

In 1978, astronomer James Christy discovered a small satellite orbiting Pluto. The moon has been given the name Charon, after the mythical ferryman who transports souls across the river Styx to the underworld. Charon is estimated to be about half the size of Pluto, making it the largest moon compared to its planet in the solar system. Pluto and Charon are sometimes called a double planet.

For five years starting in 1985, Charon was passing in front of Pluto, as seen from Earth, once each orbit. These periodic eclipses showed that Pluto, like Uranus, has its rotational axis close to the plane of its orbit. It has further been determined that Pluto is a rock-rich planet, containing between 67 and 79 percent rocky material by mass.

Observations also suggest that the planet has a polar cap of ice over at least part of its northern hemisphere. Pluto appears to be a strange new world of rock and ice. By analyzing the changing brightness of Pluto and Charon as they alternately pass in front of one another, as well as passing in and out of each other's shadows, astronomers hope to refine their estimates of the diameters, masses, and compositions of these two bodies.

With conventional rocket power, a direct flight to Pluto would take at least 10 years, probably more. No space missions are planned to Pluto in the near future. The mutual eclipses of Pluto and Charon should yield much new information, but scientists will probably have to wait until the 21st century for a close-up look.

COMETS

From the earliest times, comets have awed humanity with their brief appearances in the night sky, often accompanied by long, majestic tails stretching across the heavens. More often than not, comets were objects of fear. Many people believed they were evil omens. There was terror even in 1910, when the most famous comet of all—Halley's Comet—made an appearance.

In 1986 Halley's Comet again made its closest approach to Earth. In March of that year, unmanned spacecraft drew near to the comet, took close-up pictures, and gathered information on the comet's composition, structure, and activity.

DISTANT WANDERERS

For the majority of their lives, most comets roam in frigid darkness beyond Pluto, near the edges of the solar system. There may be as many as 200 billion comets circling the sun in a vast, spherical swarm called the

Oort Cloud. Occasionally, the gravity of a passing star flings a comet from its distant orbit to a new path in the inner solar system.

The majority of the approximately 700 comets discovered so far are long-period comets. These follow long, highly elliptical paths and take between hundreds and millions of years to complete one orbit of the sun. Over long periods of time, the gravitational tugs of massive Jupiter and other planets may change a comet's path into a more circular one. Such short-period comets complete one revolution of the sun in anything from years to a century or two. Halley's Comet is a short-period comet, since it makes one circuit approximately every 77 years. Its namesake, Edmond Halley, was the first to realize that comets can return on regular visits.

Another type of comet is the sun-grazer. Sun-grazing comets approach the sun at speeds of about 125 miles (200 kilometers) per second.

THE STRUCTURE OF A COMET

What is a comet? A comet is composed of two parts, a nucleus, or body, and a tail.

Resembling a dirty snowball, the nucleus consists of frozen gases, dust, and ice. It is an inactive chunk when far from the sun. No one has ever seen the nucleus of a comet, but some scientists think they are irregular lumps 6 miles (10 kilometers) or less in size.

When the nucleus approaches the sun, the sun's warming rays begin to cause its outer, icy layers to evaporate, releasing dust and gases. These form into a cloud, called a coma, around the nucleus.

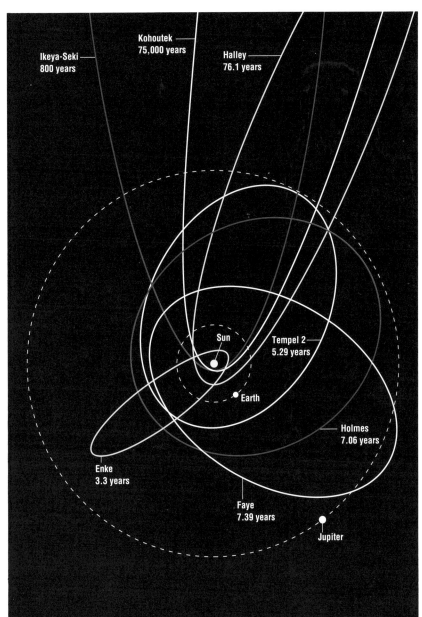

As the comet heads for its rendezvous with the sun, sunlight exerts a minute pressure on the dust, while the gas molecules become caught up in the stream of charged particles known as the solar wind, which emanates from the sun. As a result, the gas and dust form into one or more tails, always pointing away from the sun.

A comet has no light source of its own. All its light is reflected sunlight. As such, it becomes brighter the closer it comes to perihelion, which is the point on

Comets orbit the sun in a wide variety of patterns and lengths of time. Some short-period comets, such as Enke, Holmes, and Tempel 2, travel within the orbit of Jupiter. Certain long-period comets, such as Kohoutek, are not expected to return to the inner solar system for many thousands of years.

its orbit closest to the sun.

Comets are often most spectacular just after perihelion, when the tails reach their great-

This is one of the first photographs ever taken of Halley's comet. It shows the famous comet as it appeared in 1910 passing near Earth.

est length. But there are no rules—comets are almost completely unpredictable. A comet may be spectacular on one visit and unimpressive on the next. They are subject to sudden changes in brightness. Some comets disappear altogether when their supply of ice is used up.

Once the comet rounds the sun, its activity slowly ceases. The nucleus heads for the outer reaches of the solar system, millions or perhaps many billions of miles (kilometers) from the sun, before it returns for another visit.

Meteors

Often called shooting or falling stars, meteors are caused by meteoroids entering the earth's atmosphere. These chunks of metallic or stony matter can be fragments of asteroids (as discussed on page 70), fragments of comets, or the original material from which the moon and the planets were made.

Meteoroids often orbit through space in swarms. Whenever the Earth meets such a swarm, a "meteor shower" occurs. The sky seems filled with a shower of flying sparks. If the meteoroids are fragments of comets, then their swarms have orbits similar to those of comets. Whenever the Earth crosses such an orbit, meteor showers might also occur.

Meteor showers are named after the constellations from which they appear to come. Since the showers are caused by orbiting space debris, we can also plot the orbits and predict the times when they occur. For example, the Leonid meteor shower appears to come from the constellation Leo and occurs every November. We know from written records that it was seen as long ago as A.D. 902. Other important annual showers include the Quadrantid in January, the Lyrid in April, and the Perseid in August.

THE ORIGIN OF COMETS

Some astronomers believe comets were among the first things to condense from the cloud of gas and dust that spawned the sun and planets. Comets may have escaped the intense heating that wiped out the earliest history of the planets and their satellites. If so, comets could contain a record of the birth of the solar system. For that reason, scientists are anxious to learn more about them. The *CRAF* spacecraft mission is scheduled to rendezvous and fly alongside Comet Wild 2 in August, 1995, and missions to obtain a sample of cometary material and bring it to Earth are on the drawing boards.

In the meantime, scientists think they may already have microscopic bits of cometary dust in their laboratories. Cosmic dust, which falls into Earth's upper atmosphere every day, is thought to be shed by comets during their visits. Scientists have used high-flying research planes to collect cosmic dust samples for analysis. Each grain is only a few microns (millionths of a meter) across and weighs only a billionth of a gram. Hence, the grains are extremely difficult to study.

Under a powerful electron microscope, many cosmic dust grains turn out to be collections of much smaller grains joined together. This is precisely the structure scientists expect a comet to have—not hard, but loosely constructed. The spaces between individual grains may have been filled with ice.

From findings by future spacecraft and with continued, intensive observations from Earth, scientists hope to gain a better understanding of what comets are.

A typical comet away from the sun consists of a small nucleus of frozen gases, dust, and ice. As it approaches the sun, sunlight and the solar wind cause its outer layers to thaw, forming a cloudlike halo (coma) around the nucleus and one or more tails of gas and dust. As the comet swings around the sun, the tails reach their greatest length. As the comet moves away from the sun, the coma and tails begin to dissipate. The tails always point away from the sun.

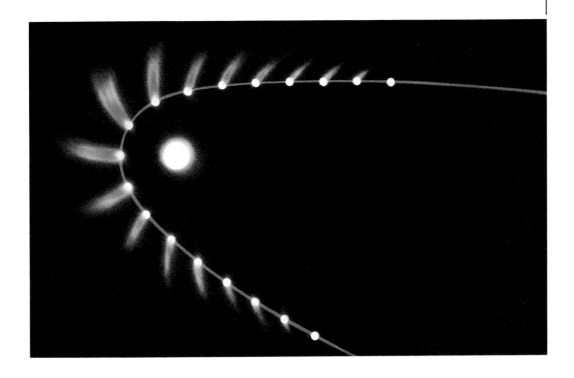

OTHER SUNS, OTHER REALMS

Beyond our solar system are immense numbers of stars, galaxies, and other objects that, along with vast stretches of empty space, make up the universe as we know it. This section of the book examines many of these objects, including the different types of stars and galaxies, nebulae, supernovae, pulsars, black holes, and quasars. Theories on the origin of the physical universe are also given, with particular emphasis on the big bang theory.

The Trifid nebula is a large ball of gas that is illuminated by stars shining within it. The Trifid is part of a large interstellar cloud located in our own Milky Way galaxy.

Galaxy M104, otherwise known as the Sombrero galaxy, displays a brightly glowing central bulge and a flat plane of interstellar dust and light. Once the galaxy's center, it is about 40 million light-years from Earth.

This color-enhanced optical photograph of Galaxy NGC 1097 shows spiral arms made up of millions of stars radiating out from the galaxy's center.

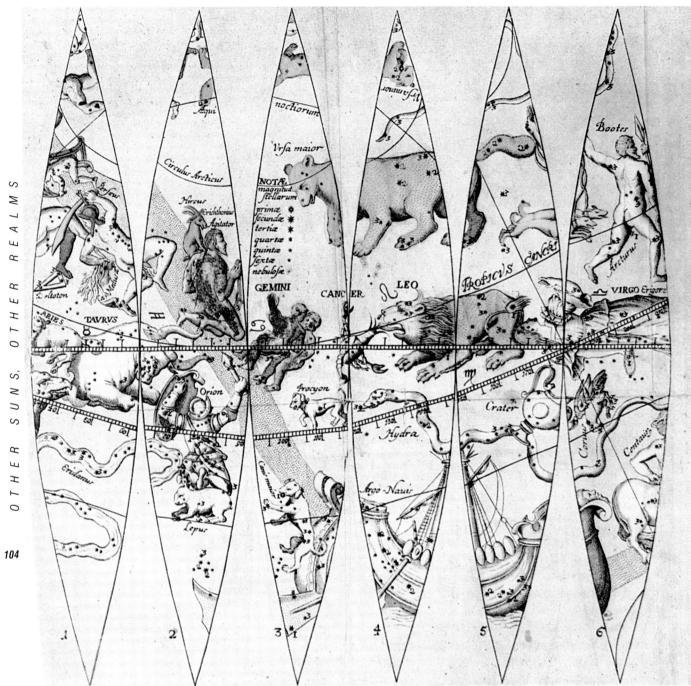

CONSTELLATIONS

Have you ever spent an evening watching the stars? To look at these tiny pinpoints of light, it is nearly impossible to imagine that they are mighty suns. The gap between appearance and reality is too great.

Recall how the ancients saw the universe: a dome sky over the flat Earth. Since this is how the sky actually does appear, astronomers still find it useful to imagine a sky dome arching over Earth—at least when mapping the stars or aiming a telescope.

The sky dome visible at any one time from any one location is only half of all space. The other half is below the horizon. So, astronomers think not in

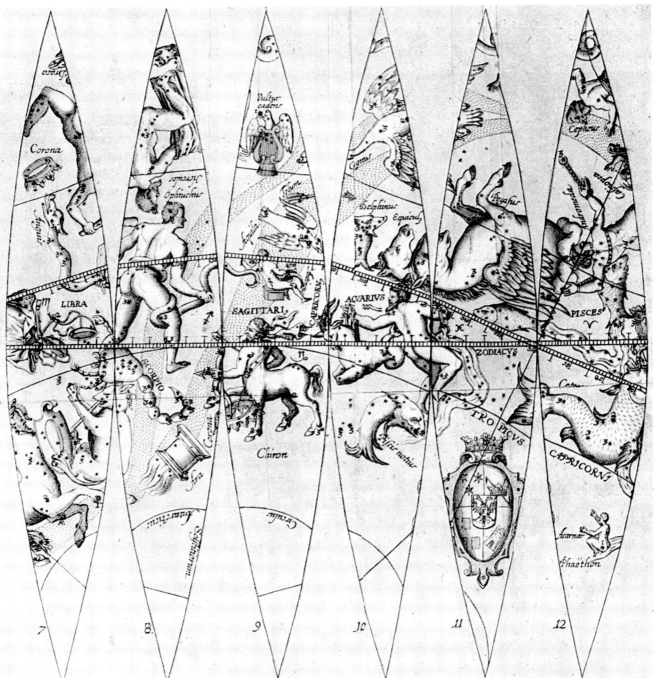

This seventeenth-century Czech drawing is a fanciful rendition of the constellations of the zodiac and some of the other major star groups. The heavier, wavy line in the center is the ecliptic, the apparent annual path the sun makes through the zodiac, as seen from Earth.

terms of a dome or hemisphere, but of a complete celestial sphere encircling Earth. The stars seem to be attached to its inside surface; we're at its center.

We know the celestial sphere is imaginary. It is only the apparent sky, not the real universe of endless space. But since this is the face the universe presents to our view, let's review some aspects of the sky as it appears from Earth.

PATTERNS IN THE SKY

From earliest times, people looking at the sky saw the stars in patterns—a triangle here, a zig-zag line there, a more complex group somewhere else. The same star patterns appeared night after night, year after year. The ancient shepherds and nomads who

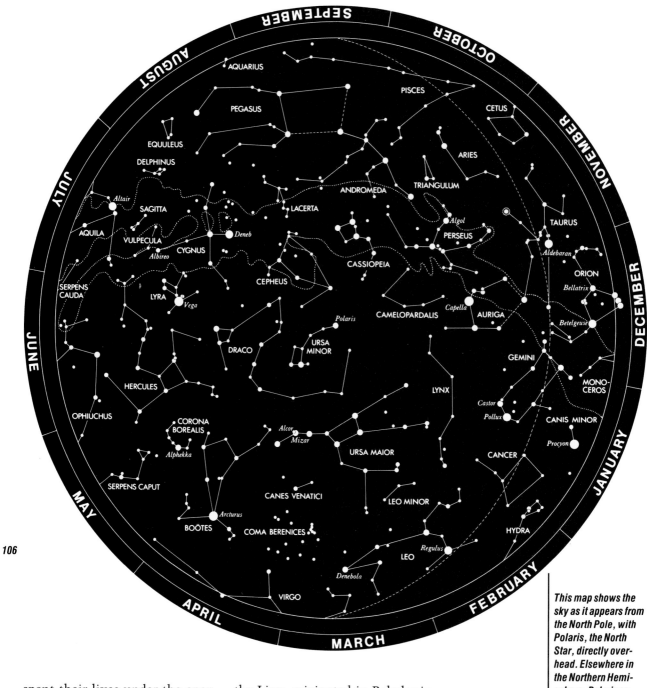

spent their lives under the open sky must have become intimately familiar with them.

These people saw the star patterns in terms of things they knew: animals, mythical characters, and various objects. In this way, the constellations were invented. The word *constellation* simply means "star group." Many familiar constellations, such as Taurus the Bull and Leo

the Lion, originated in Babylonia around 3000 B.C. or even earlier. They are among the oldest examples of ancient culture that have lasted into modern times.

To fill the holes between the ancient constellations, astronomers of more recent centuries invented some new ones. Still more were invented when European sailors first explored the Southern Hemisphere and saw a part of the sky never visible from northern lands. Today, the entire celestial sphere is divided

This map shows the sky as it appears from the North Pole, with Polaris, the North Star, directly overhead. Elsewhere in the Northern Hemisphere, Polaris appears lower in the sky. To use the map, face south and turn the map so that the current month appears at the bottom. The stars in the bottom two-thirds of the map will be visible at some time of the night.

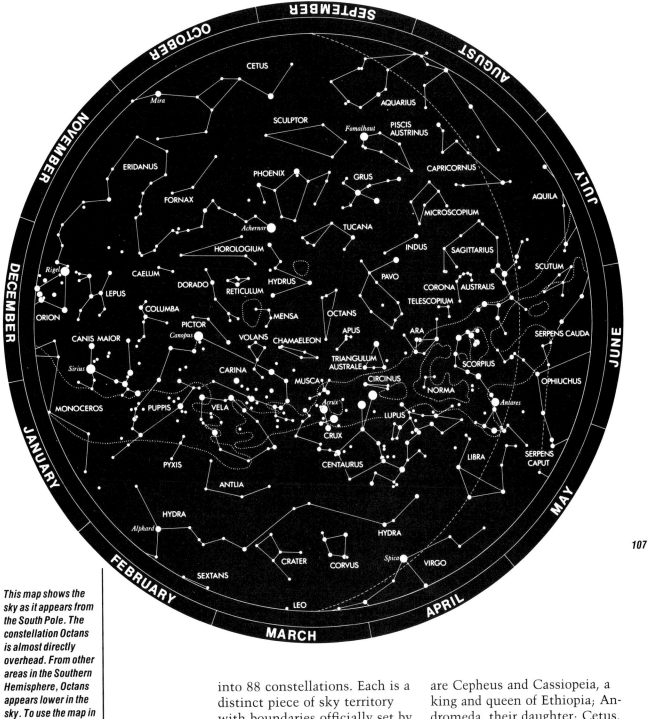

This map shows the sky as it appears from the South Pole. The constellation Octans is almost directly overhead. From other areas in the Southern Hemisphere, Octans appears lower in the sky. To use the map in the Southern Hemisphere, you would face north and turn the map so that the current month appears at the bottom.

into 88 constellations. Each is a distinct piece of sky territory with boundaries officially set by the International Astronomical Union, an international association of astronomers that is the controlling body of world astronomy.

The 88 constellations fill the sky with an odd jumble of creatures, mechanical objects, and mythical persons. Among them are Cepheus and Cassiopeia, a king and queen of Ethiopia; Andromeda, their daughter; Cetus, a sea monster the gods sent to devour her; and Perseus, her rescuer. Hercules, the legendary Greek hero, has a place in the sky. There is a ship, a sailor's sextant, a pair of scales, a clock, a telescope, and an air pump. Among the animals memorialized in the stars are two bears, a giraffe, a crab, two sea serpents, several dogs, a wolf, a crow, a rabbit, a lizard, and a fly.

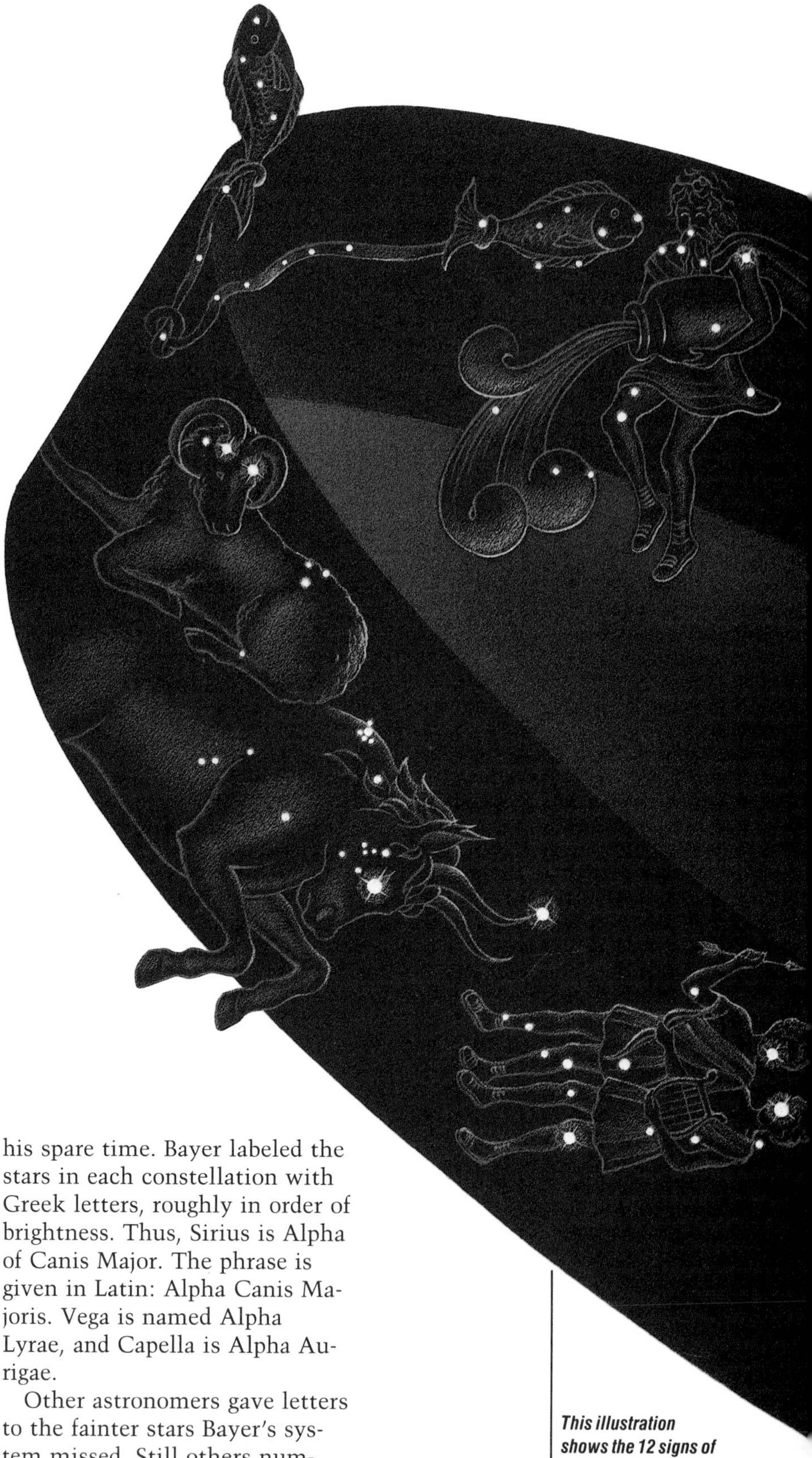

Learning your way around the brighter constellations will give you a good practical grasp of the sky. But don't expect to see actual likenesses of lions and air pumps. Whatever was going on in the ancients' minds when they named the star patterns we'll never know. Perhaps they connected the stars to form stick figures suggesting the things named. The inventors of more modern constellations, including the machines, made no claim at all of trying to match names to shapes.

Once you know some of the star patterns, you'll be able to point out bright stars and planets by name. This is the first step to becoming an amateur astronomer. After all, you can't use a telescope until you know where to point it, and the constellations make up the map of the heavens.

108

STAR NAMES

The most prominent stars have names of their own. The brightest in the sky is Sirius in the constellation Canis Major, the Large Dog. Among the brightest stars visible from our north temperate latitudes are Arcturus in Boötes the Herdsman, Vega in Lyra the Harp, Capella in Auriga the Charioteer, and Rigel in Orion.

About 3,000 stars are visible to the naked eye on a clear, dark night. Few people could remember 3,000 separate names for them. Instead, astronomers use a system that was invented in 1603 by Johann Bayer, a German lawyer who mapped the sky in his spare time. Bayer labeled the stars in each constellation with Greek letters, roughly in order of brightness. Thus, Sirius is Alpha of Canis Major. The phrase is given in Latin: Alpha Canis Majoris. Vega is named Alpha Lyrae, and Capella is Alpha Aurigae.

Other astronomers gave letters to the fainter stars Bayer's system missed. Still others numbered the stars in each constellation. These systems are quite convenient because they tell in what sky territory a star is located. Beta Cygni and 61 Cygni, for instance, are both in the constellation of Cygnus, the Swan.

This illustration shows the 12 signs of the zodiac. Note that the star patterns bear no real resemblance to the figures they portray. The zodiacal signs are used today primarily to identify different parts of the night sky.

It should be stressed that the constellation patterns are only apparent, not real. When you see a few stars forming an apparent pattern in the sky, they probably are not at all like that pattern in three-dimensional space. One star can be many times farther away than another that seems right next to it; they're merely lined up in our line of sight. The four stars forming the Great Square of Pegasus, for example, are at distances of 70, 100, 180, and 500 light-years, so they actually have nothing to do with each other. A light-year is the distance that light travels in one year. Since light travels at the speed of 186,282 miles per second, one light-year is approximately 6 trillion miles.

STAR BRIGHTNESSES

The ancient Greek astronomers called the brightest stars first magnitude, meaning "largest."

Slightly fainter ones they called second magnitude, and so on down to the faintest that can be seen, which they called sixth magnitude.

In modern times, this scale has been extended both to fainter and brighter objects. When the telescope was invented, it showed stars too dim to see with the naked eye; these were named seventh magnitude, eighth magnitude, and so on. Eventually, the system was put on an exact mathematical basis, so star brightnesses could be measured accurately. Today, a pair of binoculars will show stars of eighth or ninth magnitude, an average amateur telescope reveals stars of twelfth magnitude, and the largest telescopes in the world are able to detect stars down to about twenty-fifth magnitude.

At the same time, it was realized that some first-magnitude stars are much brighter than others. So the scale was extended upward to zero magnitude, for stars like Vega and Capella, and into negative numbers. The planet Venus gets as bright as magnitude -4.7, the full moon is magnitude -12, and the sun is magnitude -27. This system of magnitudes is encountered in every branch of astronomy.

MOTIONS OF THE SKY

As the hours of the night wear on, stars set in the west and new ones rise in the east. We know it is Earth that is turning, not the stars. But it looks as if the celestial sphere rotates around us while we stand still.

If we are viewing the heavens from Earth's Northern Hemisphere, the entire sky appears to pivot on what is called the north celestial pole. This is the point in the sky directly above the Earth's North Pole. Very near this point is a famous star, Polaris, otherwise known as the North Star. It's not very bright (second magnitude), but it isn't hard to find, because the two stars at the end of the Big Dipper's bowl point almost directly at it.

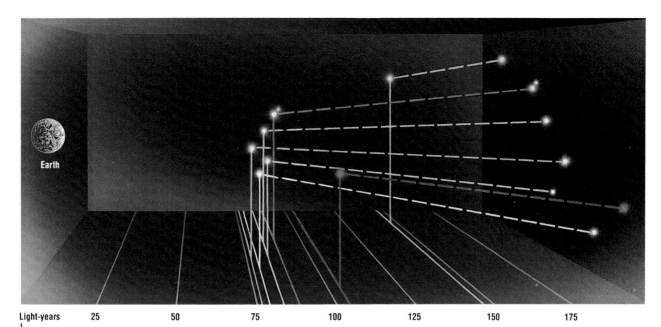

Stars in the sky all appear to be the same distance from us, but they are not. This diagram shows the Big Dipper as we see it (right) and the actual distances of the stars (center) from Earth.

The objects of our solar system—the sun, moon, and planets—appear to move among the stars. The sun appears to circle the celestial sphere once a year. This appearance, of course, is caused by Earth orbiting around the sun. As a result, we see different constellations in the night sky—that is, in directions away from the sun—at different times of the year. The Great Square of Pegasus is in the evening sky of autumn; Sirius is a winter star; Vega lights the evenings of spring and summer.

The sun's apparent path across the sky is called the ecliptic. The moon and planets also follow the ecliptic fairly closely. This well-traveled path crosses 12 major constellations known as the zodiac. These are the constellations in which the sun, moon, and planets are almost always found.

Thousands of years ago, when the planets were associated with gods or spirits or looked on as some sort of occult objects, they were assumed to influence human affairs. The study of astrology was the attempt to chart this influence, which was supposed to depend on the positions of the sun, moon, and planets in relation to the constellations of the zodiac.

Today, we know the planets and stars are physical objects and are much too far away to have any effect on human affairs. In case there was any doubt, careful studies have searched thoroughly for the slightest match-up between people's astrological charts and their lives or personalities. There turns out to be no connection at all.

On the other hand, the truth about the stars is far stranger and more amazing than anything the medieval astrologers could have dreamed up. It is to these distant suns that we now turn.

111

STARS

A star is a gigantic ball of hot gas, so hot it glows. Our sun is a very average example, as stars go. Some are much larger than the sun, some much smaller; some are many times brighter, some far dimmer. The sun falls about midway between the extremes; in fact, it is a little on the small and dim side.

Even so, the sun is a stupendous object 865,000 miles (1,384,000 kilometers) in diameter—wider than over 100 Earths. Its surface temperature is about

Stars are the most common objects in the universe—more than 200 billion billion. M3, pictured here, is a globular cluster containing millions of stars, most of which are concentrated in the center of the cluster.

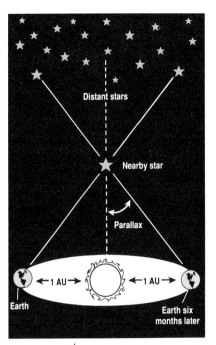

10,000° F. (5,500° C), hotter than almost any flame that can be created on Earth. Using carefully protected and filtered telescopes, astronomers find that the sun's surface is an extremely violent, boiling sea of hot gases marked by vast upheavals, stormlike activities, and explosions many times larger than Earth itself.

Yet compared to most of the stars visible to the naked eye, the sun is small. Sirius is quite a bit hotter and larger and shines with 30 times more light. Vega is 50 times brighter than the sun; Rigel puts forth 15,500 times more light.

Stars are made up mostly of hydrogen and helium. This is the key to where they get their energy. Deep inside a star, the temperature may be tens of millions of degrees; at the sun's core, the temperature is over 27,000,000° F. (15,000,000° C). This is hot enough to set off a nuclear reaction that combines hydrogen nuclei to form helium. This is the reaction that powers a hydrogen bomb. A star is like a giant, continuous H-bomb that uses its nuclear fuel steadily rather than exploding all at once.

In the 1800's, before nuclear energy was discovered, the source of the sun's power was a great puzzle. No one had any idea what could keep the sun burning for so long. If the sun were made of coal, astronomers calculated that it could shine at its present brightness for only a few thousand years before burning up completely. If it were slowly shrinking, it could gain heat from gravitational energy, but only enough for 100 million years. Geological evidence proved that Earth had been warm for much longer than this. The mystery of what powers the sun was a major spur leading to the discovery of nuclear energy. Nuclear reactions, as we shall see, are the key to understanding the lives, and eventual deaths, of all stars.

HOW HOT, HOW BIG, HOW BRIGHT?

One thing can usually be known about a star just by looking at it: its temperature. Some stars have more red in their light, such as Betelgeuse and Antares. They have a distinctly reddish tint even at a casual glance. Other stars have more orange in their light, some, like Capella, more yellow. The sun and Polaris are both white. Others such as Vega and Rigel have more blue in their light. This is the same sequence of colors that objects on Earth—from an electric stove coil to a light-bulb filament—go through when heated enough.

Stellar parallax can be used to determine the distance to a nearby star. A nearby star's position against a background of distant stars seems to shift as the star is viewed from opposite ends of Earth's orbit. One-half of the angle formed by the "lines of vision" is known as the stellar (or trigonometric) parallax. The radius of Earth's orbit is one Astronomical Unit (AU), 93 million miles (149 million kilometers). Knowing the parallax and the radius of Earth's orbit, trigonometry is used to calculate the distance to the stars.

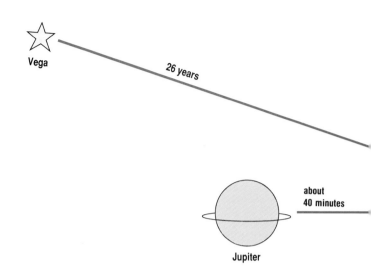

Vega

26 years

about
40 minutes

Jupiter

2,200,000 years

Andromeda Galaxy

What is a Light-Year?

A light-year is the distance light travels in a year at a speed of 186,282 miles (298,051 kilometers) per second. This distance is 5.88 trillion miles (9.41 trillion kilometers). Light reflected from the moon takes about 1.25 seconds to reach Earth. The sun's rays travel for about 8.33 minutes before they reach us. Light reflected from Jupiter takes about 40 minutes to reach Earth. The light emitted from Vega takes 26 years to travel the trillions of miles separating this nearby star from our solar system. The light we see from the Andromeda galaxy, the nearest major galaxy to our own, is already 2.2 million years old when it reaches us.

But most facts about a star cannot be known just by looking. Among these are its distance and true brightness. Sirius and Rigel appear not far apart in the winter sky. How can Sirius look brighter if it is only 23 times as luminous as the sun while Rigel is 15,500 times as luminous? The answer is that Sirius is only 8.8 light-years away, a very near neighbor as stars go, while Rigel is about 900 light-years distant. The light we see from Sirius left it just 8.8 years ago. The light that strikes our eyes today from Rigel has been traveling through space for so long that it was only a little more than halfway here when Columbus first landed in the Americas.

The most luminous stars blaze several hundred thousand times the sun's light. At the other extreme, the dimmest stars emit

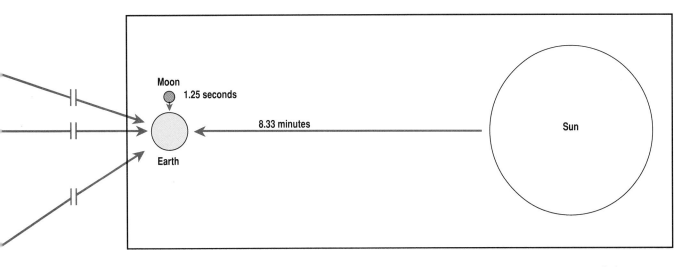

Moon
1.25 seconds

8.33 minutes

Earth

Sun

less than one one-hundred thousandth the light of the sun. If such a feeble star were to replace our sun, we would receive little more light from it than we now get reflected off a full moon.

The size of a star is another quantity impossible to know just by looking. Stars are so far away that they appear as pinpoints in any telescope, no matter how powerful. But they have almost as extreme a range of sizes as brightnesses. The biggest may be 1 billion miles (1.6 billion kilometers) or more across. It would take light fully an hour and a half just to cross the face of such a monstrous object; our sun, by comparison, is only 4.6 light-seconds across.

Other stars are as small as a few thousand miles in diameter,

and in the most extreme cases, less than 10 miles (16 kilometers) in diameter. Such an object might fit within the borders of your hometown. These are the neutron stars, extraordinary objects made of extremely dense nuclear material. They have densities, gravities, hardnesses, and magnetic field strengths beyond imagining.

Clearly, the word *star* covers an enormously broad range of different objects. If you put the brightest and faintest ones next to each other, the difference would be greater than that between a high-power searchlight and a tiny spark of glowing ash. The difference in size between the smallest and largest would be the difference between a speck of dust and a gas balloon 1 mile (1.6 kilometers) across.

However, stars are much more alike in one very important respect: the actual amount of material they contain. The least massive ones have about one-

twentieth the mass of the sun, the largest only 100 times the sun's material.

STAR CLASSIFICATION

When all this variety began to be discovered among stars, astronomers developed a way to categorize them and find the order beneath the confusion. They placed the stars on a chart called the Hertzsprung-Russell diagram, which is named after the two astronomers who invented it. Stars are arranged by temperature from left to right, and by true brightness from top to bottom. The chart has been a basic tool for categorizing and understanding stars since its invention in the early 1900's.

Those stars that lie at the top of the diagram—the extremely luminous stars—are called supergiants; hot blue ones are toward the left, cooler red ones toward the right. Below these are ordinary giants. Another variety of stars, the largest group, falls on a curving band from upper left to lower right. This is called the main sequence. These are normal, hydrogen-burning stars like our sun—which can be found near the middle of the band.

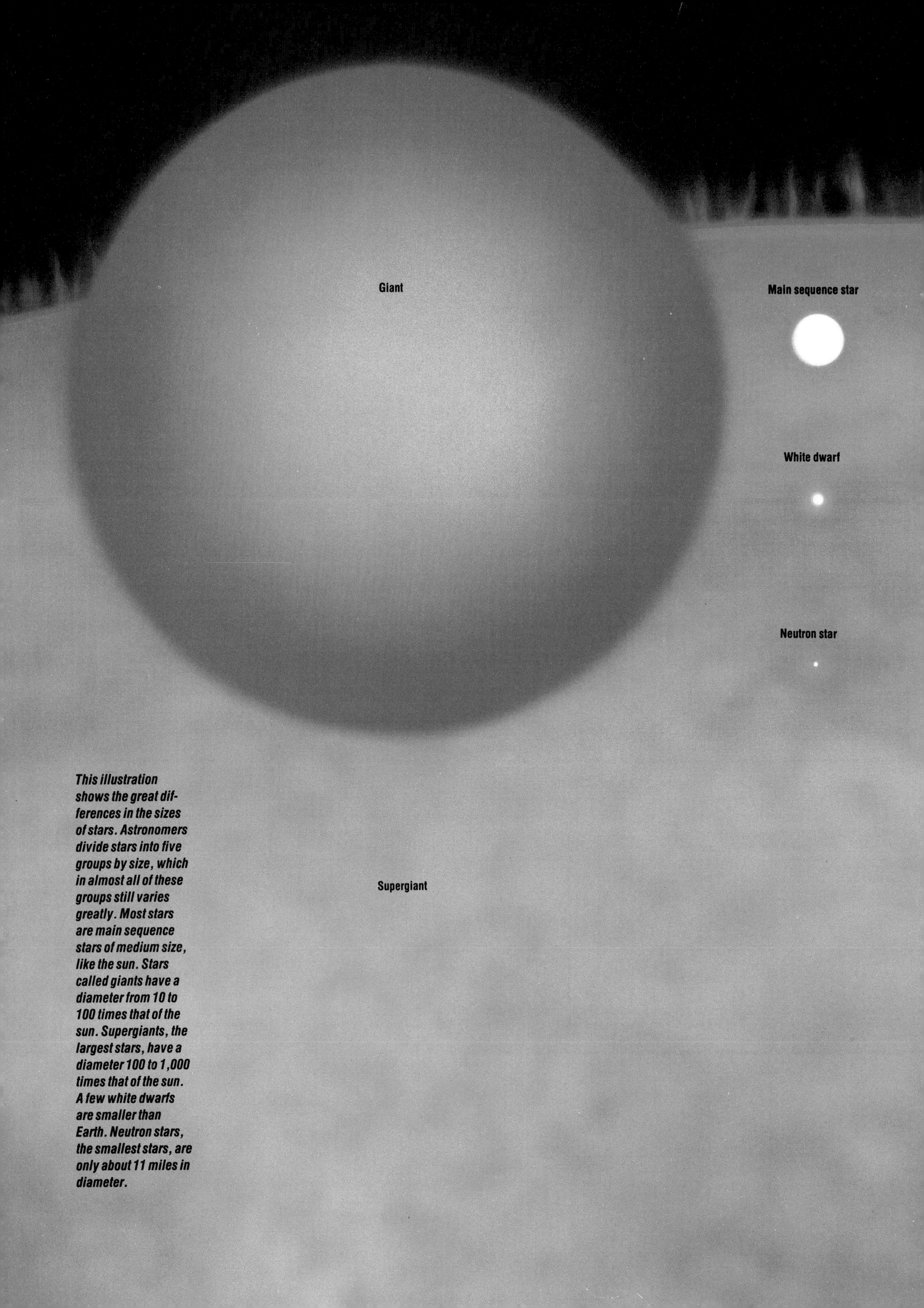

Giant

Main sequence star

White dwarf

Neutron star

Supergiant

This illustration
shows the great dif-
ferences in the sizes
of stars. Astronomers
divide stars into five
groups by size, which
in almost all of these
groups still varies
greatly. Most stars
are main sequence
stars of medium size,
like the sun. Stars
called giants have a
diameter from 10 to
100 times that of the
sun. Supergiants, the
largest stars, have a
diameter 100 to 1,000
times that of the sun.
A few white dwarfs
are smaller than
Earth. Neutron stars,
the smallest stars, are
only about 11 miles in
diameter.

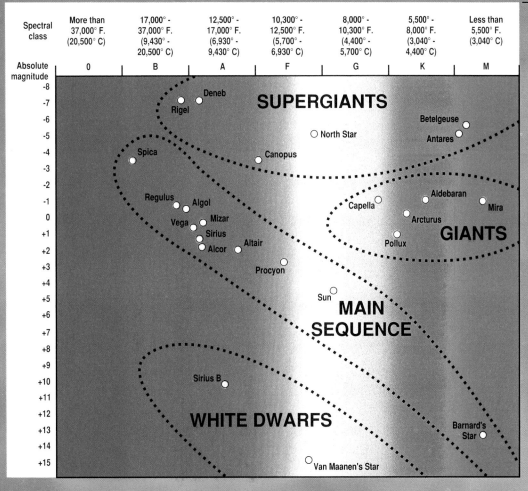

Spectral class	More than 37,000° F. (20,500° C)	17,000° - 37,000° F. (9,430° - 20,500° C)	12,500° - 17,000° F. (6,930° - 9,430° C)	10,300° - 12,500° F. (5,700° - 6,930° C)	8,000° - 10,300° F. (4,400° - 5,700° C)	5,500° - 8,000° F. (3,040° - 4,400° C)	Less than 5,500° F. (3,040° C)

Absolute magnitude

| 0 | B | A | F | G | K | M |

-8
-7   Deneb
-6   Rigel   SUPERGIANTS
-5   ○ North Star   Betelgeuse
-4   Spica   Antares
-3   Canopus
-2
-1   Regulus   Aldebaran
0   Algol   Capella   Mira
+1   Vega Mizar   Arcturus   GIANTS
+2   Sirius   Pollux
      Alcor Altair
+3   Procyon
+4
+5   Sun   MAIN
+6   SEQUENCE
+7
+8
+9
+10   Sirius B
+11
+12
+13   WHITE DWARFS   Barnard's Star
+14
+15   ○ Van Maanen's Star

The Hertzsprung-Russell diagram helps astronomers classify and study stars. The color bands illustrate spectral classes identified by the letters and temperatures across the top of the diagram. A star is represented on the diagram by a dot located horizontally according to the star's spectral class and vertically according to its absolute magnitude. For example, the sun belongs to spectral class G and has an absolute magnitude of +5. This combination puts the sun on the main sequence.

Main-sequence stars are called dwarfs to distinguish them from giants; thus the sun is a yellow dwarf. Among the dimmest stars of all are the red dwarfs, like Barnard's Star, at the bottom of the main sequence. White dwarfs are a peculiar class off by themselves. Neutron stars are beyond the diagram's lower right edge. We will return to all of these types as we follow the evolution of a typical star from its birth through its youth, middle age, and eventual death.

Notice the capital letters across the top of the diagram. These are spectral classes, which

A planetarium theater has special facilities for presenting programs about the stars and the solar system. The large projector in the middle of the theater creates an image of the sky on the domed ceiling.

is the most widely used way of sorting stars into distinctive groups. Among other things, a star's spectral type tells its temperature very accurately.

A star's light can be broken up into its different colors, or wavelengths, by sending it through a prism—the same way a prism will break up a ray of sunlight into rainbow colors. The array of colors is called a spectrum. A star's spectrum is marked by many fine lines at different wavelengths. These are called spectral lines, and they are a star's individual fingerprint. Each chemical element or compound causes its own set of spectral lines at different temperatures. So, by analyzing the lines, astronomers can tell a star's temperature and chemical composition.

Spectral types form a smooth sequence from the hottest to coolest stars. The order of this spectral sequence, as can be seen on the diagram, is O B A F G K M, with type O stars being the hottest blue-white ones and type M those that are only red-hot. Our sun is type G.

DOUBLES AND MULTIPLES

Some stars that look single to the naked eye are seen in a telescope actually to be close pairs. The first of these double stars to be discovered was Mizar, the middle star in the handle of the Big Dipper. Other prominent doubles are Albireo (Beta Cygni), Castor (Alpha Geminorum), and Epsilon Lyrae. These and hundreds more are lovely sights in a small telescope.

The two stars of a double are truly close together in space, bound together by their gravity. They orbit each other the same way planets orbit the sun.

Doubles come in almost every possible combination of star types. And they have all possible distances apart. Some pairs are so close that the stars almost touch; these orbit each other very rapidly, as quickly as once an hour. Others are farther apart and orbit more slowly. Some pairs, separated by almost a light-year, take millions of years to complete one orbit around each other.

119

The closest doubles cannot be separated with the best telescopes. Astronomers instead find them by indirect techniques. When all doubles are tallied, they actually turn out to be slightly more common than single stars.

In addition, some stars are triple, quadruple, quintuple, and up. It is fascinating to figure out what the days would be like on a planet in a multiple star system. Such planets probably do exist. There might be worlds where a sun like ours is rising in the east while a red-giant sun blazes down from high overhead and a tiny, brilliant blue sun is setting in the west—making it dawn, noon, and evening all at once.

VARIABLE STARS

The sun, luckily for us, keeps putting out the same steady amount of light and heat day after day, year after year. But certain other stars vary in brightness, often dramatically. They are known as variable stars.

One type of variable star isn't truly variable at all. Sometimes a close double is oriented so that every time the pair circle each other, one star eclipses, or blocks, our view of the other. So we see the total light of the system diminish and then return to normal once per orbit. These are called eclipsing binary stars. The most famous example, and the first discovered, is Algol (Beta Persei). It loses two-thirds of its brightness every 2 days and 21 hours. This eclipsing of its light lasts about five hours.

Another type of binary star is the X-ray binary star, which comprises a normal star and a collapsed companion circling each other. The companion star emits the X rays that first alerted astronomers to this binary star system. The intensely energetic emission appears to arise when gas from the normal star falls toward its binary companion.

Other variable stars are single stars that really do change their light output. They come in almost countless varieties. One class is the pulsating variable stars. An example of this type is the red giant Mira in the constellation Cetus; its changes in brightness can be followed with the naked eye. Pulsating variables change radically in brightness within a period of a few hours to many years. Cepheid variables, another type of pulsating variable, pulsate with shorter periods usually lasting about one week.

Another class is the exploding variables. These are stars that burst from time to time. Some undergo a relatively small outburst—increasing just a few magnitudes in brightness—every few days or weeks. Others, the recurring novae, explode more vigorously every few decades. The true novae are those that have only been seen to explode once, and with still greater violence. Even they, however, are believed to survive the event to explode again every few thousand years.

Only the outer layer of a star is involved in a nova explosion; the rest of the star remains intact. A supernova, however, destroys a star entirely. This is by far the most spectacular stellar event. A supernova is seen in our galaxy only rarely—the last one was seen in 1604—but it can become the brightest star in the sky for several weeks and can even sometimes be seen in broad daylight.

Many other kinds of variables enliven the sky. Some stars shine at their normal brightness for months or years at a time and then suddenly fade without warning. And there are stars called irregular variables that follow no clear pattern at all.

Amateur astronomers have contributed a great deal to this branch of astronomy by monitoring the variations of hundreds of stars for decades. Professional astronomers, in turn, have found the study of variables to be especially valuable for probing the inner workings of all stars and for helping to figure out their life histories from birth to death.

Friedrich Wilhelm Bessel (1784–1846), a German astronomer and mathematician, was the first person to measure the distance to a star other than the sun.

121

NEBULAE AND THE BIRTH OF STARS

So-called empty space is not totally empty. Even between the stars there is an extremely thin trace of gas, mostly hydrogen. It averages about one atom per cubic centimeter. This is still far closer to a perfect vacuum than the best vacuum that can be made on Earth.

Mixed in with the thin gas are traces of dust particles. Each microscopic particle is approximately 300 feet (900 meters) from the next. But the volume of space is so huge that these traces

A nebula is a cloud of dust particles and gases in space. The emission nebula (lower left) is brightly lit up by stars from within it. The Horsehead nebula (right) is so filled with dust and gas, that any starlight from behind it is blotted out.

orescent lamp. Or the dust particles can reflect starlight directly, forming a reflection nebula.

A dark nebula is another type. It doesn't glow at all and can be seen silhouetted against a starry background. Some dark nebulae are visible to the naked eye. If you have a very dark, rural sky, you can see the band of the Milky Way crossing overhead in the summer; it appears spotted with patchy dark nebulae blotting out the light of stars behind them.

In some nebulae, many bright stars are mixed in with the clouds. They are there by no coincidence. Nebulae are where all stars are born.

STAR BIRTH

If a cloud of interstellar matter is dense enough, it starts to contract under the force of its own gravity. Every bit of matter in it attracts every other bit, with the result that all parts of the cloud start falling together. The smaller the cloud gets, the faster it shrinks. Material falling in from different sides collides in the middle and remains there, and more material rains down

of interstellar matter add up to an important amount.

In some places, the interstellar gas and dust is a bit thicker than in others. If a relatively thick cloud happens to lie near a star, the cloud is lit up, forming a bright nebula. A well-known example is in the constellation Orion. A pair of binoculars is all that is needed to see the Great Nebula of Orion. It looks like a hazy, multi-colored mist surrounding the middle stars in Orion's sword.

A nebula (Latin for "cloud") can shine in two ways. The gas can fluoresce, or glow, under the influence of a star's ultraviolet light, much like the gas in a flu-

Sixteen hundred light-years from Earth, stars are continuing to form in the Great Nebula of Orion. The reddish streaks are huge quantities of energized hydrogen gas.

124

These two photographs show what may be new stars forming in a dust-and-gas cloud. The photograph on the left, taken in 1947, shows three stars in the cloud. The photograph on the right, taken in 1954, shows two outgrowths on two of the stars. These outgrowths may be new stars forming from the ''parent'' stars.

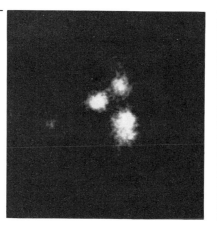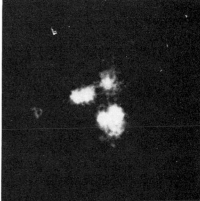

onto it. This mass at the center of the cloud heats up under the impact of all the infalling material. At this point, the object is called a protostar.

As the protostar becomes denser and hotter, pressure builds up at its center and slows its shrinking. Eventually, the heat and pressure in the center become great enough to ignite the nuclear reaction that converts hydrogen to helium. This nuclear activity provides abundant heat—enough to halt the star from shrinking any further. It shines brilliantly, and any of the surrounding cloud that remains in the vicinity is dispersed.

The star has now arrived on the main sequence of the Hertzsprung-Russell diagram. Here it will remain for much of its life, slowly using up its hydrogen fuel.

Any planets that orbit a star are formed at the same time. If the original cloud had any internal motions or rotation—which is very likely—then some of the material will form a thick disk of gas and dust orbiting the newborn star. Within this disk, smaller masses of gas and dust are expected to collapse and become planets. This is presumably how our own solar system formed. Today, we see such disks surrounding a number of stars that are in the process of being born. We could be witnessing future solar systems in the making.

STAR CLUSTERS

Stars are generally born in great bunches rather than singly. A cloud that starts collapsing may contain enough matter for many thousands of stars. Indeed, a nebula often contains many groups of stars, called star clusters.

As time goes on, the nebula is consumed to form more stars or is blown away. Left behind is one or more clusters. Dozens of such clusters are visible in binoculars or a small telescope. The most famous is the Pleiades in the constellation Taurus. It is easily visible to the naked eye in the winter sky as a little glittering cloud. Long-exposure photographs of the Pleiades reveal the last wisps of the dusty nebula from which its stars were born.

After many millions of years, a cluster is likely to break up entirely, with each star going its own separate way. That is why many stars, including the sun, eventually travel alone.

Star Formation

A star begins forming from a collapsing cloud of gas and dust (left). As the cloud continues to collapse, the central area begins to heat up, forming a protostar (center). The central area eventually becomes hot enough to ignite a continuous nuclear reaction. The collapsing ceases, and the result is a main-sequence star.

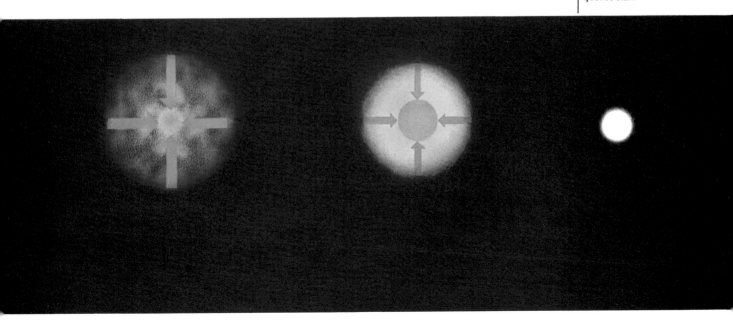

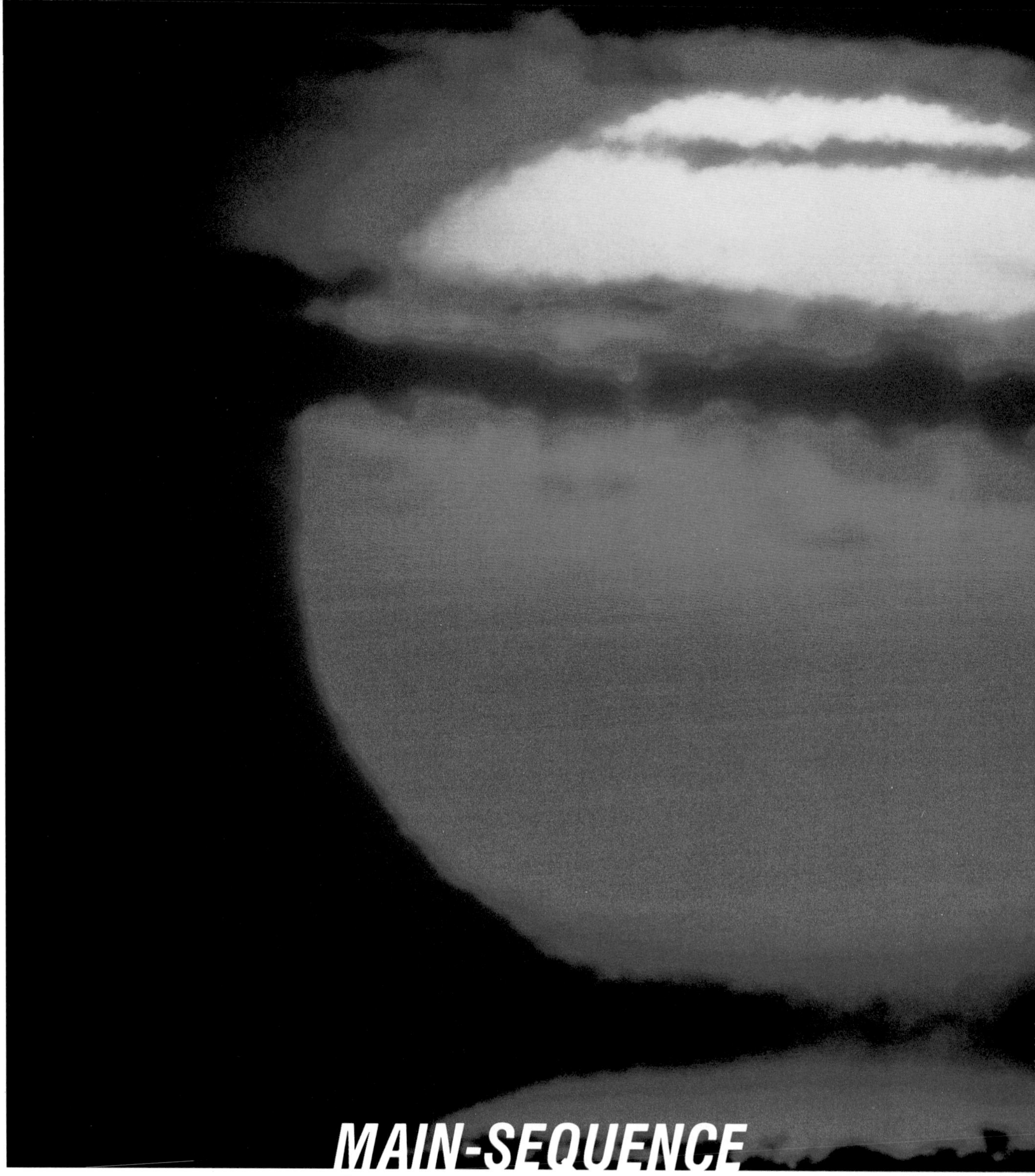

MAIN-SEQUENCE STARS

Once a star is born, it stays at the same brightness and temperature for much of its life. If it were to shrink any further, the pressure and temperature at its core would rise, causing the nuclear reaction to occur at a faster rate and create more energy,

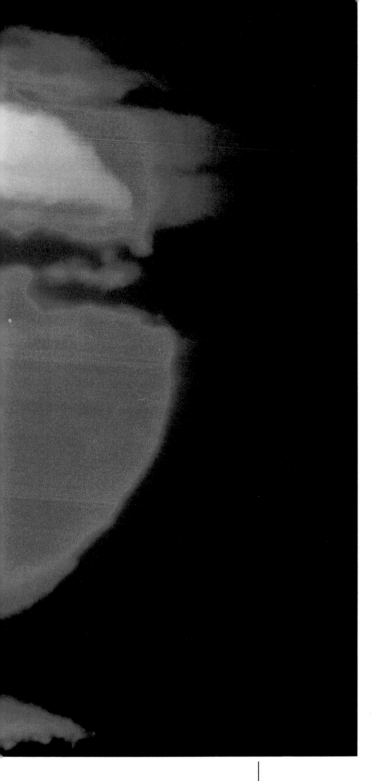

Our sun is a main-se-quence star. It re-leases energy at a steady rate and will remain about the same size for most of its life.

which would expand the star again. If something caused it to swell, it would cool and the re-action would slow, releasing less energy and allowing the star to contract again under the squeeze of its gravity. So it does not change either way. Any system in such a situation is said to be in equilibrium. The sun is a good example.

Recall the Hertzsprung-Russell diagram. The band of the main sequence represents stars that haven't changed much since birth. Note the enormous range in brightness. Some are many times brighter than the sun, and some are many times fainter. What causes some stars to burn so brightly and others to glow so dimly?

The answer is simply their mass. A star that is formed with a lot of material—30 times the sun's mass, let's say—has in-tense gravity pulling it together, so the pressure at its core is high. The nuclear reaction that converts hydrogen to helium then proceeds very fast. A tre-mendous amount of energy is re-leased. The result is a brilliant, blue-white star at the top of the main sequence. Examples are the stars of Orion's Belt.

A star with just a couple of times the sun's mass will be far-ther up the main sequence from the sun. Sirius is an example. Astronomers call the mass of the sun one solar mass. A star with only one-twelfth of a solar mass burns its hydrogen so slowly that it is a feeble red dwarf at the bottom of the main se-quence. Barnard's Star is an ex-ample of a red dwarf.

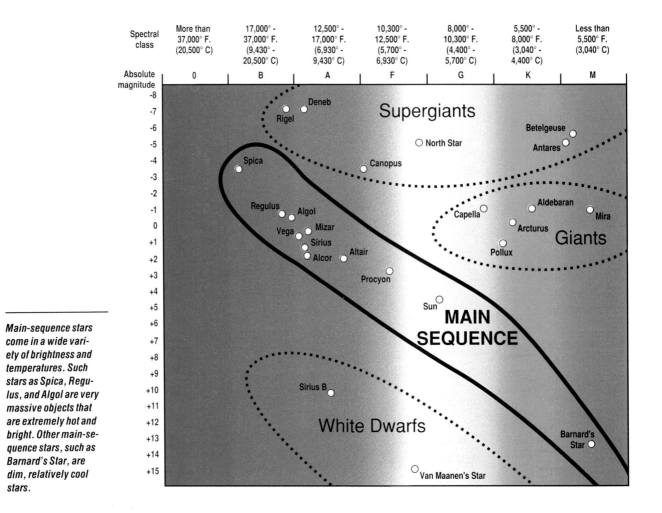

Spectral class	More than 37,000° F. (20,500° C)	17,000° - 37,000° F. (9,430° - 20,500° C)	12,500° - 17,000° F. (6,930° - 9,430° C)	10,300° - 12,500° F. (5,700° - 6,930° C)	8,000° - 10,300° F. (4,400° - 5,700° C)	5,500° - 8,000° F. (3,040° - 4,400° C)	Less than 5,500° F. (3,040° C)
	O	B	A	F	G	K	M

Main-sequence stars come in a wide variety of brightness and temperatures. Such stars as Spica, Regulus, and Algol are very massive objects that are extremely hot and bright. Other main-sequence stars, such as Barnard's Star, are dim, relatively cool stars.

The faster a star burns, the sooner its fuel will be used up. The brightest stars are using up their fuel at a very fast pace. Even though they have more fuel to use, they burn it at such a fast rate that they last only a few million years.

The sun, using its hydrogen more prudently, has a lifetime on the main sequence of about 9 or 10 billion years. Red dwarfs will outlast all other types of stars in the universe, shining on in their modest way for hundreds of billions of years. In the inconceivably remote future, there may be a time when no main-sequence stars remain but red dwarfs.

THE LIFE OF THE SUN AND EARTH

Where are we in the life of our own star? The sun, Earth, and the rest of the solar system are 4.6 billion years old. That is how long it has been since the sun and planets condensed out of a collapsing cloud of interstellar gas and dust. We know approximately how much hydrogen fuel the sun contains, and we know about how fast it is being used up. So we can estimate how much longer the sun will shine. The answer is reassuring: the sun will light and warm Earth steadily for another 5 billion years.

After that time, the sun will swell to become a red giant star and Earth will be doomed. But this ultimate fate of Earth is of no immediate concern to the human race. It is too far away.

To grasp the time involved in the lives of the sun and Earth, let's consider a scale model. We can represent the 4.6-billion-year age of the solar system by the height of the World Trade Center in New York City; it is a skyscraper 110 stories—1,350 feet (405 meters)—tall. On this scale, a million years is just 3.5 inches (8.75 centimeters). Earth and the sun formed at ground level. Most early multicellular forms of life—mud worms and the like—didn't evolve until about 600 million years ago, or around the 95th floor. About 420 million years ago—the 100th floor—the first primitive plants and animals came out of the sea onto land.

The dinosaurs lived for a period of time represented by the 104th to the 108th floors. The earliest humans appeared about 2 million years ago, only 7 inches from the ceiling of the 110th floor. All recorded history, all civilization, lies in the last 5,000 years—the layer of paint one-fiftieth of an inch thick on the ceiling of the 110th floor.

Yet, the history that lies ahead for Earth and the sun amounts to a whole additional 110-story building placed on top of the first. Earth has a much longer future ahead of it than the time since the first grubs lived in the mud of the ocean. We may think we are the crown of creation, but the history of life on Earth is clearly just beginning. The death of the sun, so far into the inconceivable future, cannot possibly concern us.

Yet even today, we see in the sky many stars much older than the sun. We can point out stars similar to the sun that are 9 or 10 billion years old and coming to the end of their lives on the main sequence. Their hydrogen is nearly used up, and already they are beginning to swell into the bloated and monstrous forms that characterize the later parts of a star's life. To this stage we now turn.

129

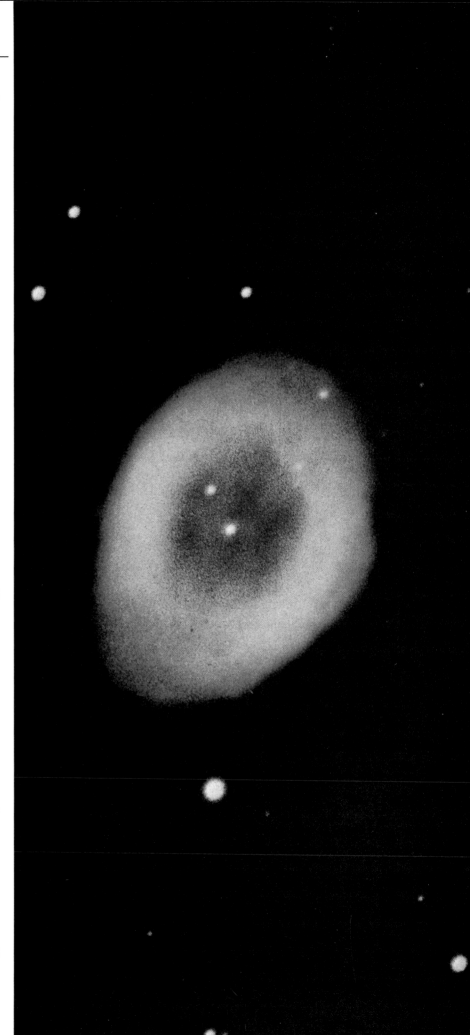

The Ring nebula, pictured here, resulted when the bright star in the middle of the nebula blew off its outer layers into space, forming a cloudlike halo of dust and gases. Such an occurrence is called a planetary nebula. It is fairly common toward the end of life for certain kinds of stars.

RED GIANTS, WHITE DWARFS

130

Nothing is forever. As ages roll on, the interior of a main-sequence star slowly begins to change. Its hydrogen fuel is used up, and helium piles up in its core. The zone where nuclear reactions take place moves from the center closer to the surface. The entire star begins to expand—slowly at first, then with increasing speed.

The star is becoming a red giant. It eventually swells to dozens or even hundreds of times its original size. Its surface cools to the point of being only

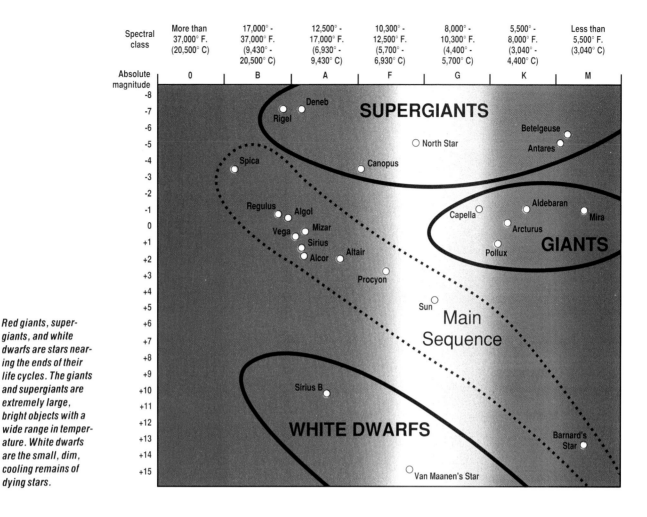

Spectral class	More than 37,000° F. (20,500° C)	17,000° - 37,000° F. (9,430° - 20,500° C)	12,500° - 17,000° F. (6,930° - 9,430° C)	10,300° - 12,500° F. (5,700° - 6,930° C)	8,000° - 10,300° F. (4,400° - 5,700° C)	5,500° - 8,000° F. (3,040° - 4,400° C)	Less than 5,500° F. (3,040° C)

Absolute magnitude

	O	B	A	F	G	K	M

SUPERGIANTS

Deneb
Rigel

Betelgeuse

North Star

Antares

Spica

Canopus

Regulus   Algol

Aldebaran

Capella

Mira

Vega   Mizar

Arcturus

GIANTS

Sirius

Pollux

Alcor   Altair

Procyon

Sun

Main Sequence

Sirius B

WHITE DWARFS

Barnard's Star

Van Maanen's Star

Red giants, supergiants, and white dwarfs are stars nearing the ends of their life cycles. The giants and supergiants are extremely large, bright objects with a wide range in temperature. White dwarfs are the small, dim, cooling remains of dying stars.

red-hot, but it has become so large that its total output of light and heat is much greater than before. Any planets nearby will be roasted and possibly melted, or they may be actually engulfed in the expanding giant.

Once the sun starts swelling, some 5 billion years from now, it will take another several hundred million years to grow into a full-blown red giant. At that time, it will be about 70 to 100 times larger than now—a gigantic fireball filling much of the sky. It will pour out 500 to 1,000 times more light than at present, heating our Earth to somewhere around 2,000° F. (1,100° C). The seas will have long since boiled away, and the atmosphere itself will be driven off into space. The world will be bare rock under a gigantic orange sun in a black sky.

Examples of stars that are already at this stage are Aldebaran in the constellation Taurus and Arcturus in Bootes.

Stars that are more massive than the sun burn brighter when on the main sequence, use up their hydrogen much faster, and expand more quickly. Red supergiants, such as Betelgeuse, are the result. These are the largest

The Death of a Star

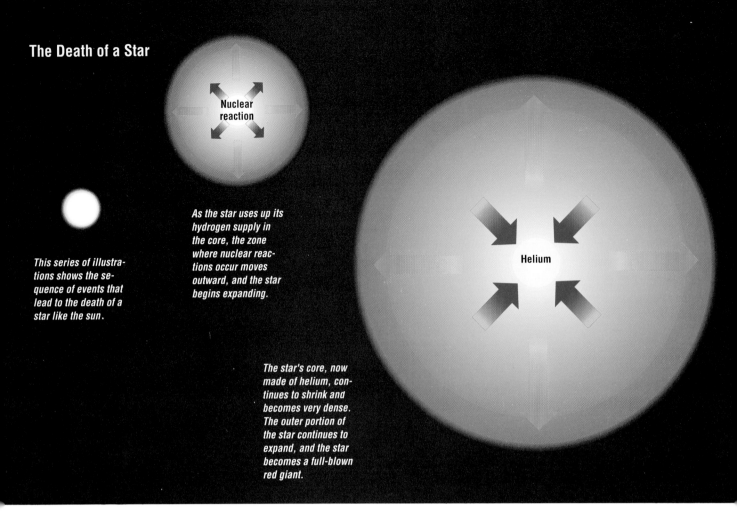

This series of illustrations shows the sequence of events that lead to the death of a star like the sun.

As the star uses up its hydrogen supply in the core, the zone where nuclear reactions occur moves outward, and the star begins expanding.

Nuclear reaction

Helium

The star's core, now made of helium, continues to shrink and becomes very dense. The outer portion of the star continues to expand, and the star becomes a full-blown red giant.

stars of all. Betelgeuse is roughly 400 million miles (640 million kilometers) across; if it were put in the place of the sun, it would fill the solar system out beyond the orbit of Mars.

LAST GASPS

A star cannot remain a red giant for very long, usually only about a tenth of the time it spent on the main sequence; it is running out of fuel. The helium core shrinks and becomes so dense and hot that another nuclear reaction starts: the fusing of helium to become carbon. This reaction provides some more energy and helps keep the star alive for a while longer. But it is shrinking and losing brilliance.

The star becomes hotter as it shrinks, moving to the left on the Hertzsprung-Russell diagram. At this stage, it is likely

to become a pulsating variable. The late stages of a star's life can be quite complex, depending on its mass. We will continue to follow what happens to one like the sun.

As time goes on, the star becomes still more unstable. Finally, a large part of its outer portions is gently blown off into space. This gas becomes a great bubble-shaped nebula around the remaining bare stellar core. This tiny core is extremely hot, so it emits great amounts of ultraviolet light, which causes the bubble to glow.

Many nebulae of this kind are visible in a telescope. They can be recognized by their roundness or symmetry and the fact that each is centered around a very

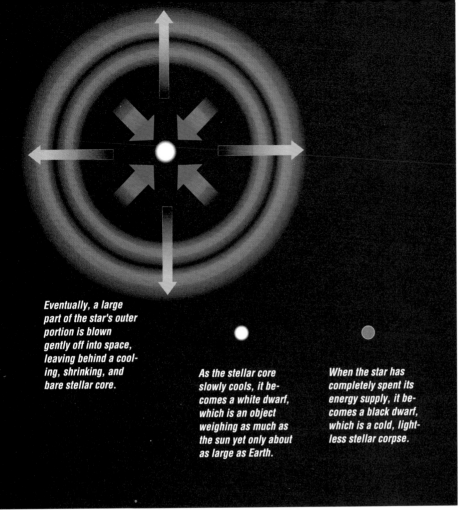

Eventually, a large part of the star's outer portion is blown gently off into space, leaving behind a cooling, shrinking, and bare stellar core.

As the stellar core slowly cools, it becomes a white dwarf, which is an object weighing as much as the sun yet only about as large as Earth.

When the star has completely spent its energy supply, it becomes a black dwarf, which is a cold, lightless stellar corpse.

hot blue star. Early observers named them planetary nebulae, because their roundness gave them a dim, ghostly resemblance to the way a planet looks in a telescope.

A planetary nebula lasts only a very short time astronomically—a few tens of thousands of years. Within this time, it disperses into space, rejoining the interstellar gas and dust out of which the star condensed so long ago.

The tiny star remaining has no more sources of energy. It slowly cools down and loses its brightness, moving to the lower portion of the Hertzsprung-Russell diagram and entering the realm of the white dwarfs.

DEAD STARS

When white dwarfs were discovered, they were almost too strange to be believed. One of the first to be found orbits the

nearby star Sirius. Astronomers in the 1800's measured the positions of stars very accurately, and they found that Sirius was wobbling back and forth by a tiny amount every 50 years. This could only be happening if Sirius were a double star, orbiting with an unseen companion.

In 1862, this companion was seen for the first time. Telescope maker Alvan Clark first spotted it while he was testing a large telescope lens. The companion became known as Sirius B. Its true brightness was found to be very slight: several hundred times dimmer than our sun. Yet its gravitational effect on Sirius showed it had to have very nearly the sun's mass.

There the matter rested until 1915, when the spectrum of Sirius B was photographed for the first time. This strange star, it turned out, is quite a bit hotter than the sun. So each square foot of its surface must be giving off more light than a square foot of the sun's surface. This leads to the conclusion that Sirius B is remarkably small. In fact, it is only about 6,500 miles (10,400 kilometers) in diameter. It is a star smaller than Earth.

To pack the sun's mass into such a small volume, Sirius B must be made of extremely dense material, thousands of times denser than lead. Its average density works out to about 70 tons per cubic inch (4.2 met-

This artist's conception of the sun as seen from Earth's surface several billion years from now shows the sun expanded into a red giant. All water will boil away, and life will cease to exist.

134

ric tons per cubic centimeter). On Earth, a single teaspoonful of its matter would weigh more than a heavy truck.

The very idea of such material at first seemed impossible. Yet it had to be true. Other white dwarf stars were soon found. One of them, named 40 Eridani B, is easy to see in a small amateur telescope.

Understanding white dwarf material requires knowing how matter is put together. In ordinary objects, atoms are packed close together and cannot be squeezed any more. This is why ordinary solids and liquids are incompressible. But an atom itself is mostly empty space.

A very lightweight shell of electrons determines an atom's size, but its tiny nucleus contains most of its mass. A scale model of an atom would be an empty gymnasium with a fly in the middle—the gymnasium being the electron shell and the fly being the heavy nucleus.

Inside a white dwarf, material is so compressed by gravity that the atoms themselves are crushed. In our scale model, the gymnasium-sized electron shells are broken up, and the nuclei—the flies—lie much closer together. That is how so much mass can fit in a small volume.

The strength of gravity on a white dwarf is tremendous. The gravity at the surface of Sirius B is about 500,000 times stronger than on Earth. If you weigh 150 pounds (68 kilograms) here, you would weigh about 38,000 tons (34,200 metric tons) on Sirius B and would instantly be squashed by your own weight into a thin film on the surface. Since a cu-

bic inch (16 cubic centimeters) of Sirius B would weigh 70 tons (63 metric tons) on Earth, it must weigh 35 million tons (31.5 million metric tons) where it actually is. On top of all this, many white dwarfs have magnetic fields thousands of times stronger than the most powerful magnets on Earth.

When the sun becomes a white dwarf, it will shine with about one-hundredth the brightness it has now. Earth, after having been roasted into a bare hulk of rock during the sun's red giant phase, will now be about $-300°$ F. ($-184°$ C) and getting colder. The dying sun will be a dazzling white pinpoint in a black sky.

For perhaps billions of years more, the white dwarf sun will slowly cool and dim, eventually becoming a cold, lightless black dwarf. This is the ultimate corpse of a star like the sun, the way it will remain forever after. Earth will be in perpetual night, lit only by starlight. Such is the fate that lies ahead for the solar system, in ages beyond imagining.

135

SUPERNOVAE AND NEUTRON STARS

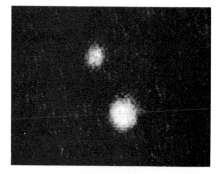

The Crab nebula in Taurus is the remnant of a star that exploded into a supernova nearly 1,000 years ago.

Some stars are born with less than a few times the sun's mass and end relatively quietly as white dwarfs. But a more massive star has a more eventful career. During its supergiant stage, its core may suddenly collapse and explode, blowing the whole star to pieces with stupendous violence. A star undergoing this fate is known as a supernova. It shines with a billion or more times the sun's light for several weeks.

Deep within the Crab nebula is a pulsar, which is believed to be the remaining core of the original star. The pulsar's light varies from bright (top) to virtually invisible (bottom) in just a few thousandths of a second.

In the summer of 1054, such an exploding star appeared in the constellation Taurus. For more than three weeks, it was so bright it could be seen in broad daylight, according to the Chinese astronomers who watched it. In the same spot today is the famous Crab nebula, so named because it looked vaguely crablike through telescopes in the 1800's. This nebula is the debris of the stellar explosion that happened over 900 years ago. It is still expanding outward from the explosion site at a speed of several thousand miles per second.

In 1989, astronomers detected the remains of two supernovae witnessed relatively recently by Earth-bound observers. It was the first time that astronomers had seen both a supernova explosion and the resulting supernova remnant. One remnant was of a star that exploded in 1957; the second was the remnant of a star that exploded in 1885.

In the center of the Crab nebula is the remainder of the supernova's core. It is an example of the strangest star of all: the pulsar.

Pulsars were first discovered with radio telescopes. In 1967, something in the constellation Vulpecula was found to be emitting regular pulses of radio energy every 1.3 seconds. Nothing known at the time could account for this. The discoverers thought at first they might be picking up signals from another civilization. But pulsars proved to be natural phenomena. Many more were discovered, including the one in the Crab nebula. That one pulses at an especially rapid rate: 30 times a second.

Pulsars are spinning neutron stars. A neutron star is an extremely dense and tiny object, about 10 to 12 miles (16 to 19 kilometers) in diameter. In this small volume is compressed an entire star's mass. A neutron star is rather like a white dwarf carried to ridiculous extremes. The atomic nuclei of its matter are packed very tightly together. Most of them are converted into a fluid of pure neutrons. Hence, a neutron star is mostly liquid, with a solid crust.

The solid crust gives neutron stars a structure somewhat like that of a planet, with a surface temperature of about 180,000° F. (100,000° C). There is also evidence of quakes occurring in these stars, though we would have to call them starquakes instead of earthquakes. Like a planet, neutron stars might have mountains on them, perhaps an inch high. This would mean that the half-inch high atmosphere would be extended into space.

One teaspoon of neutron star matter weighs several billion tons. If all of humanity were compressed to the same density as that of a neutron star, we would occupy the same volume as a drop of water. If the entire Earth were compressed to the same density as that of a neutron star, it would have a diameter of 100 yards, the same as the length of a football field. The Earth could then sit on the playing field of most major league ball parks, looking like a giant scoop of ice cream. However, the gravitational force on it would be so great that it would bore a hole into the center of the Earth.

The force of gravity on the surface of a neutron star is 100,000 million times greater than the pull of gravity on the Earth's surface. The spinning magnetic field is also 100,000 million times greater than the magnetic field of the Earth or the sun. This spinning field accelerates nuclear particles to unheard-of energy levels, causing the observed radio emission previously mentioned and creating the emitting particles of cosmic rays.

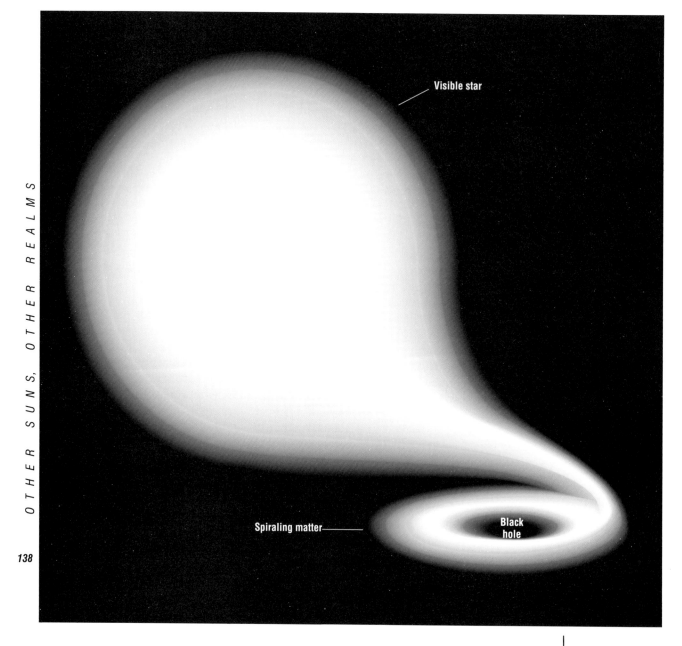

Visible star

Spiraling matter

Black hole

BLACK HOLES

Other than becoming a white dwarf or a neutron star, there is a third possible end to a star's life—a black hole. This is the strangest object of all.

The basic concept of a black hole isn't hard to grasp. Every astronomical body has a certain escape velocity. This is the speed at which an object has to travel in order to escape the body's gravity and enter into space. Earth's escape velocity is 7 miles (11 kilometers) per second. An object traveling upward with less than this speed will fall back down to Earth, while one traveling faster will fly away into space.

The sun's escape velocity is 386 miles (618 kilometers) per second, since the sun is more massive and has stronger gravity

The illustration above shows how a black hole might be detected by observing its effect on a visible star. Matter from the star would spiral into the black hole at high speeds. This would produce temperatures hot enough to give off X rays that could be detected by orbiting X-ray telescopes.

than Earth. The escape velocity from a neutron star can be as high as 150,000 miles (240,000 kilometers) per second, which is getting close to the speed of light.

What if the escape velocity is greater than the speed of light? Then even a light beam aimed up from the star will curl over and fall back down. No light will escape. To an outside observer, the body would look like a black void.

This much was recognized by astronomers as early as 200 years ago. But the true nature of a black hole had to await Albert Einstein's general theory of relativity. Einstein found that space and time are intimately related, and that gravity consists of a curvature or distortion of four-dimensional space time—something impossible to visualize, but true nonetheless. A black hole is a region with gravity so intense it has removed itself from the rest of space. Nothing is left behind but an opening. With the exception of gamma rays and X rays, nothing ever leaves a black hole.

Do such things really exist? The answer is quite possibly yes. If a dead star has a mass greater than about three times the sun's,

it cannot become either a white dwarf or a neutron star. Instead, its gravity is so strong it will keep collapsing indefinitely. As it shrinks, its gravity intensifies, its escape velocity surpasses the speed of light, and it becomes a black hole. From then on, we can see nothing more of it. The hole formed by a star of 10 times the sun's mass will be about 38 miles (61 kilometers) across.

As far as anyone knows, the collapsing star inside might go on shrinking to become a true mathematical point, with zero size and infinite density. Or perhaps at extreme densities, new laws of physics, about which we know nothing, take over. There is reason to think the inside of a rotating black hole may be a gateway to whole other realms of space time and perhaps other universes. However, it is certain that a human being who either entered or fell into a black hole would be torn apart.

Some black holes may already have been detected. A few double stars have been found in which an invisible member of the pair gives very strong evidence of being a black hole. In addition, much larger black holes with masses of millions or billions of suns might exist at the centers of many galaxies, including the Milky Way. The question is far from settled, but most astronomers assume the universe does indeed contain black holes in great numbers.

GALAXIES

Arching across the night sky among the stars is a dim, hazy band of light. The Romans called it the *Via Lactea*, meaning the Milky Road or Milky Way. According to one of their legends, it was milk from the goddess Juno. Aside from that, they had no idea what it could be.

Only in the early 1600's, when Galileo turned his telescope on the Milky Way, did its true nature become clear. Its pale glow, Galileo saw, is made up entirely of very faint stars—swarms and clouds of them "so numerous," he wrote, "as to be almost beyond belief."

Questions immediately arose: Why are so many stars gathered in this narrow band? Why aren't they scattered evenly across the sky?

In the 1700's, English astronomer William Herschel made a

*This artist's concep-
tion of our Milky Way
galaxy shows the spi-
ral arms radiating out
from the bright central
hub. Our solar system
is located in one of
these arms.*

serious effort to find out. Using his large telescopes, he counted the numbers of faint stars in different places around the sky in hopes of estimating their true distribution in space. He concluded that the universe of stars is arranged in a great, irregular, flat slab, with Earth and the sun deep inside it.

Herschel was not far wrong. Our Milky Way galaxy (the word *galaxy* comes from a Greek word meaning "milk") is indeed a very flat, pancake-shaped swarm of stars with a bulge in the center. We are inside it. We see more stars when we look through its breadth than when we look through its thin sides. This is why the Milky Way seems to ring Earth in a circle. Picture a blueberry's viewpoint inside a pancake; this is similar to our visual perspective of the Milky Way galaxy from Earth.

But the galaxy is much larger than Herschel and other early astronomers thought. They didn't know about interstellar dust that pervades the galaxy and in many places blocks our view beyond a few thousand light-years. Today, we can penetrate through the dust with radio telescopes and other means, and so we have gained a more complete picture of our galaxy.

ANATOMY OF THE MILKY WAY

We know our galaxy to be an enormous, flat pinwheel of stars, gas, and dust, indeed about as flat as a pancake. The Milky Way is almost 100,000 light-years across and just a few thousand light-years thick. Its bright central bulge, or nucleus, is somewhat thicker, perhaps 10,000 light-years through.

The sun is located about 30,000 light-years from our galaxy's center, near the inner edge of one of the spiral arms. Most of the stars we see in the sky belong to our small galactic neighborhood. The bright center lies

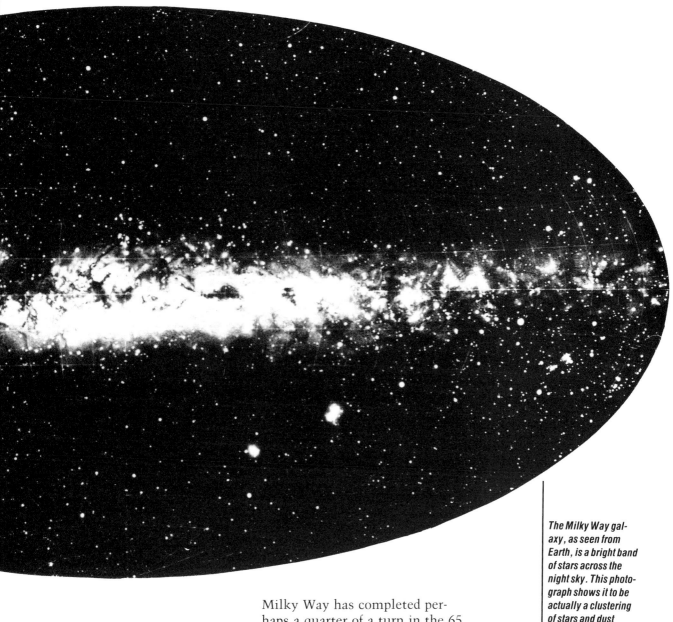

The Milky Way gal-axy, as seen from Earth, is a bright band of stars across the night sky. This photograph shows it to be actually a clustering of stars and dust clouds, strung out un-evenly across the night sky.

in the direction of the constellation Sagittarius, and it is hidden from us by dust.

The Milky Way galaxy contains hundreds of billions of stars and extensive clouds of gas and dust. Its spiral arms trail streamers of billions of suns and thousands of bright and dark nebulae. Like all spiral-shaped galaxies, the Milky Way is slowly rotating. The motion is so slow that no change whatever can be seen during the course of human history. The sun, like most of the stars near us, orbits around the galaxy's center once every 250 million years. The

Milky Way has completed perhaps a quarter of a turn in the 65 million years since the dinosaurs died out on Earth.

The galaxy's nucleus is an extraordinary place. Here the stars are much more abundant than in our own thinly populated region in one of the arms. Any planets in the central zone must have skies packed with brilliant stars.

In the center of the nucleus is a mysterious, very small area that emits powerful radio, infrared, and other radiations. It may well involve a massive black hole, whose gravitational force provides the energy which produces these strange phenomena. Scientists theorize that the

gravitational force accelerates to enormous speeds any interstellar gas falling into a black hole. The gas would then collide with other gases moving on different trajectories. The immense energy of such motion would be converted into heat and then into radiation.

In the outermost parts of the

This large cluster of galaxies, located in the constellation Virgo, is about 60 million light-years from the Milky Way galaxy.

galaxy, stars become few and far between. A very large but dim halo of such thinly scattered stars surrounds the Milky Way and other galaxies out to great distances. The halo around the Milky Way is not flat like the disk but is almost spherical. One component of the halo is fairly prominent: the globular clusters, which are dense balls of up to hundreds of thousands of stars. Dozens of globular clusters may be seen with a small telescope, some with a pair of binoculars.

DISCOVERY OF OTHER GALAXIES

The nearest spiral galaxy beyond the Milky Way was long thought to be the Andromeda nebula. It is a fuzzy patch dimly visible to the naked eye in the evening sky of autumn. In even the best telescopes, it looks like a gaseous, spiral-shaped nebula.

Several thousand other such spiral-shaped nebulae had been cataloged by 1900. Twenty years later, photography was revealing hundreds of thousands of these objects. A debate raged among astronomers about what they are. Some believed the spiral nebulae were just small gas clouds relatively nearby. In this view, the Milky Way was the entire universe. Other astronomers argued that they were whole other Milky Ways extremely far away—separate island universes millions of light-years distant.

By the mid-1920's, the island universe theory was winning. Final proof came when photographs taken with large new telescopes resolved the hazy glow of various nebulae into swarms of extremely faint stars.

We now know that the Andromeda galaxy is 2.2 million light-years away and no longer our closest galactic neighbor. The Large Magellanic Cloud is only about 160,000 light-years away. The Small Magellanic Cloud is about 180,000 light-years away. These galaxies (first recognized as such in the early 1900's) look like hazy patches of light. They are named after Ferdinand Magellan, the Portuguese explorer who in the early 1500's recorded that they looked like clouds. Other galaxies are much farther away. In fact, scientists have discovered a group of faint blue galaxies so far away that they could be the first such objects to have formed in the universe. The galaxies are so distant that astronomers believe them to be about one billion light-years from the edge of the universe. A large amateur telescope can see certain galaxies as far away as 200 or 300 million light-years.

SPIRALS, ELLIPTICALS, IRREGULARS

Galaxies come in a variety of forms. By the middle of this century, astronomers were photographing them in great numbers and trying to understand their many shapes.

A key discovery made along the way was that there are two kinds of star populations in gal-

Galaxy M51, otherwise known as the Whirlpool galaxy, shows the basic structure of a spiral galaxy.

axies. The disks of spiral galaxies contain much gas and dust, as well as many young stars. In these parts of galaxies, including our sun's own region, star formation is still proceeding briskly. The objects making up this part of a galaxy are called Population I objects.

A spiral galaxy's central zone and outer halo are quite different. Here, there is practically no gas and dust. Star formation ceased long ago, and the stars are all very old. These regions contain Population II objects.

Different galaxies have different mixes of the two populations. Some spiral galaxies have very small nuclei, or, in the extreme case of some irregular galaxies, none at all. They are made up almost entirely of Population I material. Other spirals have very big central hubs and almost no spiral arms. Extreme cases are the elliptical galaxies, which are all nucleus: pure Population II material. They have virtually no interstellar matter or young stars at all.

Galaxies were first classified along these lines by Edwin Hubble during the first half of this century. Elliptical, or type E, galaxies lack spiral structure or any detail at all. They are pure hazes of stars. (This may be because, as some scientists believe, elliptical galaxies are formed when two spiral galaxies merge, and the molecular gas in the merging galaxies virtually collapses into the center of the galaxies.) Hubble subdivided the ellipticals into types ranging from E0, for perfect spheres, to E7, for the most elongated ovals.

Spirals, type S, he classified in part by how large the nucleus is compared to the spiral arms. Type S0 is much like an elliptical, having only traces of spiral structure. In type Sa, spiral arms start to be visible; in type Sb the nucleus is smaller and the arms larger; and type Sc has thick, bright arms and a small nucleus.

The Milky Way is thought to be between type Sb and Sc.

Some spirals have a bar-shaped central hub. These barred spiral galaxies are called SBa, SBb, and SBc.

EVOLUTION OF THE MILKY WAY

The two kinds of star populations are the key to understanding much about the history of the universe. The oldest of Population II stars are almost as old as the universe itself. They are almost pure hydrogen and helium, since that is the material with which the universe began.

All the heavier elements—including most of the atoms making up Earth and our own bodies—were formed later, by nuclear reactions inside the cores of stars. When these stars came to the ends of their lives, they either exploded as supernovae or puffed off their outer layers back into space. Either way, the heavy elements they had created in their cores rejoined the interstellar gas and dust. In this way, the interstellar matter in Population I regions is being slowly enriched with atoms of elements other than hydrogen and helium.

By the time the sun and solar system formed from an interstellar gas cloud, heavy elements had been recycled back into it from earlier generations of stars. So there were enough heavy elements to form planets like Earth.

Everything around us but hydrogen and helium—every rock, tree, bit of metal, even our flesh—is made of atoms that were created inside stars that blew apart billions of years ago. We are made of the ashes of dead suns. The life cycle of the Milky Way galaxy is intimately tied up with the life of our world, and of ourselves.

This barred spiral galaxy, NGC 1530, is of the type SBb. It is in the constellation Camelopardalis.

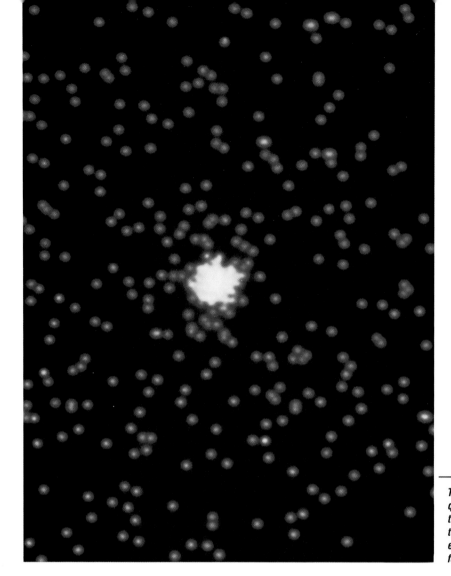

This X-ray image of a quasar was made by the Einstein Observatory, which was a satellite outfitted with four X-ray telescopes.

QUASARS

Throughout this book, we have been looking farther and farther out into space. Let's speed up the journey and extend our view to the very limits of the observable universe.

As far as can be seen in any direction beyond the Milky Way, space is sprinkled with other galaxies. They usually group together in what are known as galaxy clusters. Our Milky Way is itself part of a small cluster known as the Local Group; it is a few million light-years across.

As we look farther out to tens and hundreds of millions of light-years, the clusters themselves gather into superclusters.

Our Local Group is on the edge of the nearest supercluster, whose center is in the constellation Virgo at a distance of 30 or 40 million light-years.

The farther we look into the distance, the smaller and fainter galaxies appear. At several billion light-years, they have shrunk almost to pinpoints, barely distinguishable from faint foreground stars in the best telescopes. Yet, even at this tremendous distance, galaxies all seem to be pretty much alike. All other parts of the known universe appear to be just about the same as our own region.

ACTIVE GALACTIC NUCLEI

Beyond a billion light-years or so, however, a new and very rare type of object starts appearing. In the early 1960's, astronomers started finding brilliant, starlike objects at these distances that were putting out so much light they defied all reason. They were given the name quasi-stellar objects, or quasars. A quasar is many times more brilliant than an entire galaxy with hundreds of billions of suns. The record-holding quasar as of 1985 is 100,000 times more luminous than the entire Milky Way.

Yet, these objects must be much smaller than galaxies. They have been seen to vary in brightness, and in radio and X-ray output, as rapidly as in a few weeks or days. This must mean they are no larger than a few light-weeks or light-days in size. Otherwise, variations at different parts of the quasar would cancel out, and we would not see any changes. By comparison, recall that a galaxy is typically 100,000 light-years in size. For all its brilliance, a quasar is not much larger than the solar system.

Many astronomers believe only one known source of energy can produce so much light: the gravitational force of a very massive black hole. As material nears the black hole's edge, it becomes extremely compressed and heated even before entering the hole. This can give off more energy than even nuclear reactions could provide. A quasar is thought by some to be powered by a black hole weighing millions or billions of solar masses, located at a normal galaxy's core. We don't see the rest of this galaxy because the quasar so greatly outshines it.

In fact, weaker objects of the same kind are found in the cores of certain galaxies nearby. These active galactic nuclei are like miniature quasars. Even the Milky Way seems to have a black hole at its center, emitting still less energy—presumably because less matter is falling into it.

Still, much remains to be learned about quasars. Some astronomers believe that they are not nearly as far away as the edge of the universe, but their evidence is not strong. More powerful telescopes and other technological advances should help us learn much more about these mysterious objects in the years to come.

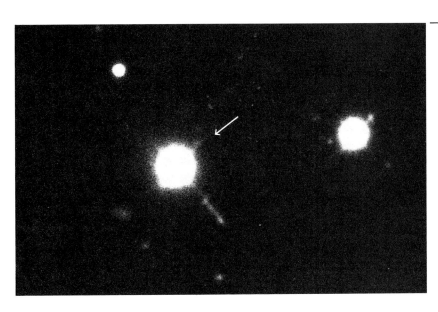

In 1963, Quasar 3C 273 became the first quasar to be identified optically. It is the largest of the objects in this photo. The jet being emitted by it contains highly energized material.

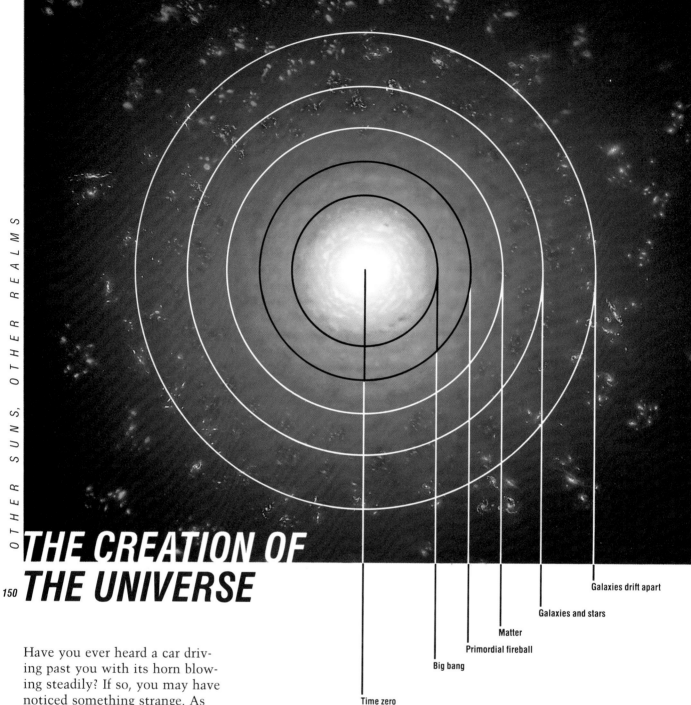

Galaxies drift apart

Galaxies and stars

Matter

Primordial fireball

Big bang

Time zero

THE CREATION OF
THE UNIVERSE

150

Have you ever heard a car driving past you with its horn blowing steadily? If so, you may have noticed something strange. As the car approached, the horn was high-pitched, but it dropped to a lower pitch as soon as it went by.

The horn itself didn't change. The driver of the car heard the same pitch all along, even while you heard it sliding down to a lower note. What you heard was an everyday example of what is known as the Doppler effect. Astronomers have turned this phenomenon into one of their most useful tools.

As the car is approaching, sound waves from the horn pile up in front of it. Thus the distance between each wave—the wavelength—becomes shorter, corresponding to a higher pitch. As the car recedes, the sound waves are spread farther out behind it, so the wavelength is longer and the pitch is lower.

The same effect occurs with light waves. A star approaching us has its light shortened in wavelength. This is called blue-shifted light because blue light

*This section discusses one widely held theory of the creation of the universe. There are many other theories of creation.

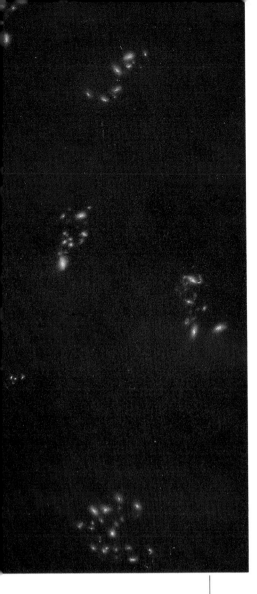

This is a "picture" of the history and future of the universe according to the big bang theory. At time zero, all forces and matter were unified. There was no time or space. The big bang was triggered and the universe was born, consisting chiefly of strong radiation, pure energy. Within a fraction of a second, the universe became a rapidly expanding primordial fireball. Atomic nuclei began forming. After about 300,000 years, the still-expanding fireball consisted mostly of matter, mainly in the form of hydrogen and some helium. After about 100 million years, matter started forming into galaxies and stars. As the universe continually expanded, the galaxies also drifted apart, moving farther and farther away from each other. The universe will continue expanding forever, the galaxies moving ever farther apart.

has the shortest waves. A receding star has its light red-shifted to longer wavelengths. By measuring the wavelengths of a star's spectral lines, its velocity toward or away from us can be found with great precision.

THE BIG BANG

In the early 1900's, astronomers began measuring the velocities of galaxies by this method. By 1929, a very strange fact was showing up. Distant galaxies are all moving away from us. The light coming from them is red-shifted. The farther a galaxy is, the faster it is fleeing away.

At first, it might seem like the Milky Way exerts some kind of repulsive power that pushes other galaxies off. But astronomers quickly realized that the situation would look just the same from anywhere else in the universe. Every galaxy, disregarding small local motions, is moving away from every other.

Imagine a balloon with dots painted on it. As the balloon is blown up, every dot gets farther from its neighbors. The galaxies are like the dots. Space itself, like the balloon, is enlarging and carrying them apart.

If the galaxies are all flying apart, then obviously they were closer together in the past. Tracing back the paths of the fleeing galaxies, astronomers realized that they all—indeed, all the matter in the universe—were packed together in a dense mass 10 to 20 billion years ago.

The universe is expanding like the debris of a gigantic explosion. Astronomers named this event the big bang. It was an amazing discovery and completely unexpected. Some astronomers resisted the idea of a big bang and looked for ways the expansion of the universe could happen without it. They invented steady state theories allowing the universe to be infi-

nitely old. But most astronomers abandoned these theories.

Strong proof for the big bang theory came in 1965 when two scientists working for Bell Laboratories were testing a sensitive microwave antenna. Microwaves are short radio waves. From all parts of the sky, the antenna picked up weak but unexplained microwave noise. This microwave background radiation is thought to have come from the big bang fireball itself. It is probably the actual white light of the big bang explosion, red-shifted all the way down into the microwave spectrum during its 10- to 20-billion-year flight through the expanding universe. New evidence, however, suggests that another event, which must have occurred shortly after the big bang, distorted or amplified the radiation.

THE FIRST MOMENTS OF EXISTENCE

The study of the universe as a whole is called cosmology. In recent years, a key part of cosmology has consisted of following the big bang back to earlier and earlier moments, ever closer to what is called time zero. Let's follow this hypothetical trail of events.

The microwave background radiation is believed to have come from the big bang fireball as it stood when the universe was only about 700,000 years old. The farther back we look, the hotter and denser the universe becomes. When the universe was only a few minutes old, scientists believe it was hot enough for nuclear reactions to fuse hydrogen into helium. The ratio of helium to hydrogen we see in the cosmos today is evidence of events that may have taken place only a few minutes from time zero.

We can follow the big bang still further back, again, hypothetically. At the first ten-thou-

151

Big bang theory

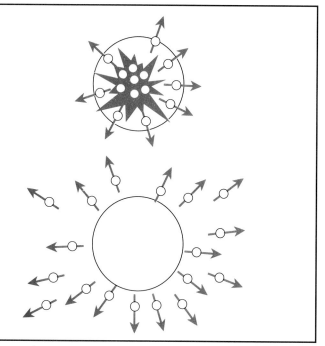

The big bang theory states that the universe began with a huge explosion (top) and that all galaxies will expand and move away from the center of the universe indefinitely (bottom).

Pulsating universe theory

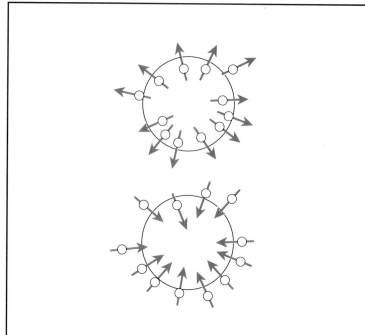

The pulsating universe theory states that all galaxies are flying apart from a previously compacted mass (top). This expansion will eventually stop, and the universe will begin to contract, due to gravity (bottom). When the universe becomes highly condensed, it will explode and begin another phase of expansion.

152 sandth of a second, the temperature of the universe was about 1.8 trillion degrees Fahrenheit (1 trillion degrees Celsius). Everything would have been a pure, uniform sea of radiation and elementary particles, mostly protons, neutrons, and electrons.

A similar situation prevailed at still earlier times and higher temperatures. Going even further back, physicists now believe that tremendously important events took place at the incredible time of 10^{-35} second after time zero. Written out, this fraction of a second would be a decimal point followed by 34 zeros and a 1. Before this time, the universe may have been hot enough to unite all the basic forces of nature except gravity into one grand unified force. The separation of these forces at this time may have given rise to the laws of physics that govern the universe now.

The going gets even more difficult here. But physicists have been tremendously excited in recent years by the way many features of the modern universe are explained by events at this incredibly early moment. The scenarios involved are called inflationary universe theories, because they require the big bang to go through very sudden enlargement, or inflation, at this time.

It may seem incredible that we can actually know what happened in such a brief moment so long ago. But, if the inflationary universe theories hold up, events at that time made the kind of universe we have today. We may thus be surrounded by evidence of what happened in the very first instant of the big bang.

THE MOMENT OF CREATION

What caused the big bang? Where did it come from? Until recently, no one had the slightest idea. Time zero seemed to be where science came to a halt. Yet, buried in the laws of atomic behavior that govern the every-

Steady state theory

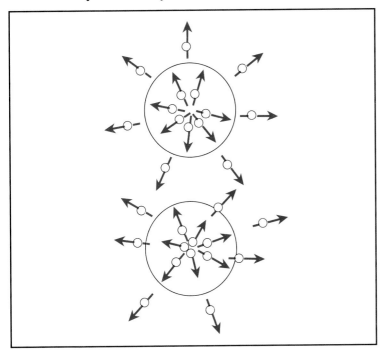

The steady state theory claims that the universe has always been expanding at a constant rate and that new matter is constantly being created (top). Hence, there is always the same amount of matter in a given space (bottom).

day world, an answer may be hidden. There are certain circumstances under which a particle of matter—an electron, say—appears out of absolutely nothing, exists for a brief instant, and vanishes back into nothing. This is seen happening in laboratory settings all the time.

The theory describing these events, if applied to the universe, would allow for larger masses to be spontaneously created out of absolute nothingness in the same way. Once this happens, the inflationary universe process can take over and cause a full-blown big bang.

MANY UNIVERSES

And there's more. All such theories predict that if a big bang can happen once, it can happen many times. There may be many other universes, resulting from many other big bangs. It is no use to ask where these other universes are. They are completely separate from our universe, our bubble of space time. If they do exist, they would probably be in some sort of different dimension we cannot break into. So how can we ever find out if other universes really exist? Perhaps we will never know.

Then again, some scientists think a piece of evidence is already at hand. There are many strange coincidences in the laws of physics governing our universe, and these coincidences are very lucky for us as living beings. No life of any kind could exist if certain physical laws were changed by a tiny amount.

It looks as if the universe were carefully hand-adjusted to allow the existence of living things.

On the other hand, these strange coincidences may be accounted for if there are indeed many universes, as these newest big bang theories predict. Most universes would lack the special coincidences allowing life. So they would remain forever barren. Only a very few may happen, purely by chance, to have the right conditions. Therefore, we might be in one of the rare universes that allows life. So even though it looks specially fine-tuned, our universe may have come about as much by chance as all the others.

THE FRONTIERS OF KNOWLEDGE

In the last few centuries, we have gone from thinking that Earth is the center of the universe to understanding that it goes around the sun; from seeing the sun as the center of everything to recognizing that it is just one star lost among the swarms of stars in the Milky Way; from thinking the Milky Way is the only galaxy to recognizing that there are many others. And now, we may be on the edge of seeing our universe, the result of our big bang, as itself being only one of many.

The realms outside and before the big bang must be the wildest, least known, and most exciting frontiers of science. Untold discoveries are just beginning to open up, and realms of which we have no conception may be explored in the lifetimes of people now living.

Index

Acknowledgments

The publishers acknowledge the following sources for illustrations. Credits read from top to bottom, left to right, on their respective pages. Charts and diagrams prepared by the WORLD BOOK staff unless otherwise noted.

Cover Jet Propulsion Laboratory (JPL). **8** Gary Ladd; The British Museum. **9** WORLD BOOK photo. **10** Anthony Miles, Bruce Coleman Inc. **12** The Metropolitan Museum of Art, New York. **14–16** Granger Collection. **17** Granger Collection; Bettmann Archive. **18–19** Granger Collection. **20** National Portrait Gallery, London; Culver. **21** Dennis Di Cicco. **24** High Altitude Observatory, National Center for Atmospheric Research; National Aeronautics and Space Administration (NASA); JPL. **25** National Optical Astronomy Observatories (NOAO); NASA. **26** NASA. **30** C. Falco, Photo Researchers. **31** Kitt Peak National Observatory. **32** High Altitude Observatory; Ned Haines, Photo Researchers. **34** Rob Wood, Stansbury, Ronsaville, Wood, Inc. **38** WORLD BOOK illustration by Anne Norcia. **40** NASA. **45** WORLD BOOK diagram by Margaret Ann Moran. **48** Mount Wilson and Palomar Observatories. **49** Bettmann Archive. **51** NASA. **52** Russ Kinne, Photo Researchers. **54** Dennis Di Cicco. **57** Photo provided by D. B. Campbell of the National Astronomy and Ionosphere Center at Cornell University, Ithaca, New York. **58** JPL. **59** JPL; Lowell Observatory. **60** NASA. **63** JPL. **64** JPL; NASA; Lowell Observatory. **66** JPL. **67** JPL. **68** Kinuko Craft. **69** American Museum of Natural History. **71** Ernest Chilson, Flagstaff Chamber of Commerce; James M. Baker. **72–80** JPL. **82** NASA. **84** William K. Hartmann; Bettmann Archive. **85** NASA; JPL. **86** WORLD BOOK illustration by Roberta Polfus. **87–92** JPL. **93** JPL. **94** JPL. **95** JPL. **96** Don Dixon; U.S. Naval Observatory. **97** Hale Observatories. **98** C. Nicollier. **100** WORLD BOOK files; WORLD BOOK illustration by Rob Wood. **102** James Sugar, Black Star; U.S. Naval Observatory; NOAO. **103** Jean Lorre, Science Photo Library. **104** Simone D. Gossner, Smithsonian Institution. **106–107** WORLD BOOK illustrations by W. J. M. Tirion. **108** WORLD BOOK illustration by Roberta Polfus. **111** WORLD BOOK illustration by Anne Norcia. **112** U.S. Naval Observatory. **116** Glenn Gustafson. **118** WORLD BOOK illustration by Ray Villard, Davis Planetarium. **121** Bettmann Archive. **122** © Royal Observatory Edinburgh (David F. Malin, Anglo-Australian Observatory). **124** © Royal Observatory Edinburgh (David F. Malin, Anglo-Australian Observatory); Lick Observatory. **125** WORLD BOOK illustration by Roberta Polfus. **126** George East. **130** California Institute of Technology. **132** WORLD BOOK illustration by JAK Associates. **134** WORLD BOOK illustration by Herb Herrick. **136** California Institute of Technology and Carnegie Institution of Washington; Lick Observatory. **137** Hale Observatories. **138** WORLD BOOK illustration by JAK Associates. **140** WORLD BOOK illustration by Anne Norcia. **142** Lund Observatory, Sweden. **144** Kitt Peak National Observatory. **145** Science Photo Library. **146** NOAO. **147** NOAO. **148** H. Tananbaum, Harvard/Smithsonian Center for Astrophysics, Cambridge, MA. **149** NOAO. **150** Don Dixon.